Digital Shakedown

Harry Halikias

Digital Shakedown

The Complete Guide to Understanding
and Combating Ransomware

 Springer

Harry Halikias
New York, NY, USA

ISBN 978-3-031-65437-4   ISBN 978-3-031-65438-1   (eBook)
https://doi.org/10.1007/978-3-031-65438-1

This Springer imprint is published by the registered company Springer Nature Switzerland AG
The registered company address is: Gewerbestrasse 11, 6330 Cham, Switzerland

If disposing of this product, please recycle the paper.

This book is dedicated to my family for their support and patience throughout the writing process.

Foreword

Over the past few years we have seen a shift in cyber-attacks, specifically a move away from one-off attacks and increasingly a number of long-term persistent campaigns conducted by nation state actors, including a record level expansion of multi-faceted extortion ransomware incidents. Financially motivated cybercrime consistently represents a serious-to-catastrophic threat to multiple critical infrastructure sectors and will likely continue to do so in the near future. A few ransomware trends have emerged: ransomware-as-a-service (RaaS) models, where attackers provide ransomware tools and services to less skilled criminals, are lowering the barrier to entry for cybercriminals and significantly reducing the time and cost required to execute attacks; extortion threat actors are continuously evolving their monetization strategies; and the ransomware ecosystem at its core is being used to monetize software vulnerabilities. Several policy levers are under consideration for thwarting these types of attacks and penalizing the perpetrators, including the development of government-led cybersecurity and resilience baseline cybersecurity standards and performance goals; mandatory disclosure requirements following an incident and ransom payments; international cooperation between the United States, its allies, law enforcement agencies, and the private sector to share cyber threat intelligence and best practices; and potential bans on ransomware payments for both public and private sector entities. Today more than ever, the public and private sectors need to collaborate more than we have in our history to ensure we're operating from a common threat picture on these extortion attacks but critically, also, considering and implementing sound policies to prevent, respond to, and mitigate damages from ransomware attacks.

Stacy O'Mara

Introduction

Ransomware is a human-controlled malicious program that locks files or computers, typically via encryption, and then demands that the system owner pay to regain access. This book examines what ransomware is, how it works, its financing, relevant laws, how to prepare and respond, and possible solutions. The book is organized and presented in five parts.

Chapter 1 documents what ransomware is and how it works by examining its history and development. In this chapter, I analyze how it fits into the current threat landscape by placing it within two industry frameworks. The analysis includes a review of the technology used to implement it and the technology it exploits. This part ends with a brief look at zero-day ransomware, highlighting a proposal by other academic researchers to detect it.

In Chapter 2, I introduce a new model for conceptualizing ransomware called the Three Cs of Ransomware: Cryptography, Cryptocurrency, and Customer Service. They represent the core components that make ransomware profitable.

Chapter 3 explores attitudes toward ransomware in a way that other scholars have not previously attempted; through conducting a survey, I uncover business and cybersecurity professionals' attitudes about paying ransoms. In addition to the business impact of the issue, I investigated prevailing legal and political concerns regarding ransomware to determine how the problem affects state and local governments. To do so, I interviewed a New York State Senator who is deeply involved in developing pioneering legislation on ransomware. New York Senate Bill S7246 would outlaw the use of public funds to pay ransoms for ransomware attacks. Senator Boyle explains how his taxpayer-focused policy position prevents taxpayer money from flowing to cybercriminals. Chapter 3 also analyses existing criminal, cybersecurity, privacy, and financial laws and regulations that victims must follow. It also includes a review of breach reporting considerations.

In Chapter 4, I look into the dichotomous situation with cyber-insurance, which leads to the question of "should you pay?". My answer to the "should you pay?" question is based on current ransomware trends and considers the different roles involved in making the decision. Deciding whether to pay is just one piece of the incident response and recovery process. The rest of Chap. 4 appraises current response and recovery challenges and makes incident preparation suggestions.

Finally, in Chapter 5, I look at possible solutions to the ransomware problem. All of them have varying degrees of effectiveness depending on the tradeoffs that societies are willing to accept.

The report is a deep dive into an ocean of information on ransomware. It will be most useful for professionals in information security, business, legal, finance, information technology, policymaking, and law enforcement to gain a greater understanding of how ransomware works and how to fight back effectively.

Acknowledgements

This book would not have been possible without many people's kindness, generosity, and wisdom. Although each bit of support may have seemed minor at the time, it helped me tremendously.

I want to acknowledge the following people for their support:

Dr. Alan Usas and Professor Shriram Krishnamurthi, thank you for believing in my ability to succeed at Brown University.

Jon Platt, William Starke, Cornel Schuler Jr., Karen Santiso, and Elicia Felix-Hughey for their support and encouragement.

Professor Linn Friedman, who was one of my advisors. She encouraged me to be ambitious with this project. Her expertise and support were invaluable.

Professor Anna Lysyanskaya, who was one of my advisers. Professor Lysyanskaya gave me entirely new ways to think about how cryptography and cryptocurrencies can be used in ransomware attacks.

Former State Senator Hon. Phil Boyle. This book would not be what it is without your support. Your legislation was ground-breaking at the time because it discouraged paying criminals using taxpayer money and encouraged spending money on security improvements. I cannot thank you enough for allowing me to interview you and telling me about the 1973 Veterans Affairs National Personnel Records Center (NPRC) fire. The story about the fire expanded my thinking about resiliency and enriched this book.

Dr. Susan Bily Lidner and Dr. Sofia Halikias-Aguilar for sharing their experiences when taking on similar projects.

Professor Tim Edgar and Professor Bernardo Palazzi for reviewing my work and sharing their ideas.

Vikas Bangia, Peter Billson, Andrea Greene-Horace, Victoria He, Mary Rose Martinez, Joshua Moorhouse. You all supported this book in multiple ways, and it wouldn't be possible without you.

Greg Foss and Rick McElroy from VMware for answering questions about their RSA Conference presentation. I am grateful for your generosity by sharing your time and knowledge. It is an example of how all cyber professionals should support each other.

Pam Greenberg from the National Conference of State Legislatures for helping me track ransomware legislation.

Stephanie Wank Kofman for reviewing a section of this book and generously sharing her expertise.

Jen Kim, Peter Levins, Asaf Rapoport, Tony Chryseliou, David Aaron, Daymon McCartney, John Lancaster, Fred Gutierrez, Jordan Schroeder, Rahul Bhardwaj. The team at Springer that guided me through the publishing process include Susan Lagerstrom-Fife, Priyadarsini K, and Kate Lazaro. All of the survey participants that shared their experiences.

Podcasters: Steve Gibson and Leo Laporte from the Security Now Podcast. Johannes Ullrich from the ISC StormCast. Douglas Brush from Cybersecurity Interviews. I listen to these podcasts regularly and they all consistently provide high quality information about the current state of cybersecurity and privacy.

And a special thank you to Stacy O'Mara for sharing her knowledge with me and writing the foreword to this book.

About the Book

Ransomware is a modern-day form of piracy and extortion in which computer files are the prized loot and companies or societies are the victims. It is an old human nature problem that occurs at internet speed, which does not require the moral decision of what to do with the hostage. However, that does not make it a victimless crime. There are countless stories of companies going out of business because they cannot pay. Boardrooms, the C-suite, business owners, information security departments, and policymakers also struggle with the complexities of an issue that seems easy (just restore from backups) but is far more complicated. The fact is that a major ransomware incident tests every aspect of an organization, from information technology to legal, finance, marketing, and even the cafeteria. This research shows how all organizations are at risk of a catastrophic incident. Ransomware is examined from every angle, from perceptions to technology, legal issues, policy-making challenges, and the high-octane fuel that is cyber insurance. It reviews how ransomware impacts various industries before answering the questions of "should you pay?". The research then examines the complexities of response and recovery. Finally, solutions to the ransomware problem are proposed. This work is a comprehensive review of ransomware's challenges that is desperately needed by those in the board room and state house alike. It provides recommendations about how to deal with the problem and where to go for help. Most importantly, it takes on one of the most complex and misunderstood problems that the internet and businesses have ever faced.

Contents

Chapter 1
What Ransomware Is and How it Works

Contents

Abstract This chapter explores the evolution, impact, and technological underpinnings of ransomware, a cybersecurity threat that has burgeoned into a multi-billion dollar concern. Originating with the 1989 AIDS Trojan, ransomware has evolved from simplistic schemes to sophisticated operations leveraging cryptocurrencies such as Bitcoin for untraceable transactions. This chapter delves into the ransomware threat landscape, illustrating its progression from data-locking to data exfiltration techniques, notably highlighted by the Maze ransomware's introduction of double extortion. Through the lens of the Lockheed Martin Cyber Kill Chain and comparisons with other malicious activities, this chapter assesses ransomware's unique position in the cyber threat hierarchy and its operational tactics, including the use of publicly available tools for significant financial gain. Furthermore, it addresses the critical challenge posed by zero-day vulnerabilities, offering insights into defense mechanisms such as anomaly monitoring and network segmentation. By synthesizing historical milestones, technical analyses, and contemporary countermeasures, this analysis provides a comprehensive overview of ransomware's trajectory and underscores the necessity of a multifaceted approach to cybersecurity in mitigating this pervasive threat.

History and Timeline of Ransomware

The first ransomware seen in the wild was the 1989 AIDS Trojan, which resembles modern-day variants. Computer users inadvertently launched the program from infected floppy disks that locked or encrypted local computer files.[1] The ransomware then demanded that victims mail $189 ($408 in 2021 dollars) to Panama in return for a decryption utility.[2] Victims could recover without paying the ransom because of cryptography implementation flaws,[3] which is also similar to some modern-day cases.

The next important milestone in ransomware history was when Adam Young and Moti Yung wrote a paper predicting ransomware in 1996 called "Cryptovirology: Extortion-Based Security Threats and Countermeasures."[4] (Ransomware was called "cryptovirus" in 1996. It evolved to "ransom-ware" in a 2005 FBI statement[5, 6] before evolving into ("ransomware.") The paper correctly foretold ransomware's utility but was not imaginative enough to predict the use of cryptocurrencies to pay ransoms. Interestingly, it proposed that ransomware would encrypt locally stored "e-money," and victims would pay to get their e-money back.[7] The reality turned out to be much more dire for victims and profitable for attackers.

The next time ransomware appeared was in 2005, when the security company Websense (now Forcepoint) helped a customer who had their files encrypted.[8] The encryption came with a ransom note demanding $200.[9] A security researcher named Joe Stewart unlocked the files without the victim paying the demand, and he astutely observed that a better-designed attack could succeed.[10] He also pointed out that the banking system was traceable, making it difficult to get away with such attacks.[11] Like Young and Yung, Stewart did not foresee the potential of electronic currencies for ransomware growth.

The invention and adoption of Bitcoin made ransomware a viable attack. Bitcoin is an electronic form of currency launched on January 3, 2009.[12] It was precisely the fuel needed for ransomware to succeed. Bitcoin transactions are global, fast,

---

[1] KnowBe4 (n.d.).

[2] Ibid.

[3] Ibid.

[4] Young and Yung (1996).

[5] Associated Press (2005a).

[6] Associated Press (2005b).

[7] Young and Yung (1996).

[8] Associated Press (2005a, b).

[9] Ibid.

[10] Ibid.

[11] Associated Press (2005a, b).

[12] Lopatto (2019).

anonymous, and difficult to trace. Threat actors launched the first cryptocurrency-based ransomware attacks by 2013.[13]

Since 2013, the use of ransomware has become a multi-billion dollar problem. Cybersecurity Ventures forecasts that ransomware damage will cost the world $20 billion in 2021.[14] That number includes ransoms, recovery costs, lost business, and other damages.

Ransomware in the Context of the Threat Landscape

Ransomware is a destructive and exploitative threat that has increasingly gained press attention because of its ability to take down entire businesses and extort victims with the threat of releasing sensitive data. Ransomware is a rising threat according to the 2020 Verizon DBIR, but it remains the third-highest source of breaches among malware and the sixth-highest "threat action" found in incidents overall.[15] It is 13th in threat actions that lead to breaches behind phishing, stolen credentials, and "other" activity. The reason ransomware is ranked higher in incidents than breaches is the change in the goals of ransomware. Threat actors originally used ransomware as a "locker" or "wiper," but the financial incentives to pay have evolved with the recent pivot from hostage-taking to blackmail (Table 1.1).

Ransomware is notorious for making money, destroying data, and shockingly large ransom payments, but that overshadows the many steps required for a ransomware attack to succeed. The best way to assess ransomware in the overall threat landscape is to compare it to the Lockheed Martin Cyber Kill Chain. The Cyber Kill Chain is a "model for identification and prevention of cyber intrusions activity. The model identifies what the adversaries must complete in order to achieve their objective."[16] It has seven steps:

1. Reconnaissance
2. Weaponization
3. Delivery
4. Exploitation
5. Installation
6. Command & Control (C2)
7. Actions on Objectives

The Kill Chain steps for ransomware can be summarized as enumerating a target's vulnerabilities to develop a malicious payload that threat actors use to exploit a system for financial gain. Most of the focus is on the "Actions on Objectives" part

---

[13] Goodin, You're infected—if you want to see your data again, pay us $300 in Bitcoins (2013).

[14] Morgan (2020).

[15] Langlois (2020).

[16] Lockheed Marten (n.d.).

Table 1.1 Ransomware news by year (2010–2020)

2010	73
2011	82
2012	417
2013	1,072
2014	1,636
2015	2,885
2016	12,014
2017	39,016
2018	17,639
2019	17,640
2020	29,598

Source: https://advance-lexis-com.revproxy.brown.edu/api/permalink/d75d0b4b-8bbc-4138-a5d5-4843df68f9b1/?context=1516831

(financial gain), when a large part of the attack occurs in the first six steps on the kill chain. The cybersecurity community talks about stopping ransomware, but ransom payments are a goal of a much larger attack. Figure 1.1 demonstrates that a wide range of techniques are required for success. The focus is on ransom payments, but the less flashy topics of configuration errors, unpatched systems, social engineering, and insider threat are a larger portion of the risk landscape.[17]

Ransomware vs. Other Destructive Attacks

The MITRE ATT&CK framework has identified a long list of destructive attacks.[18] The most famous are relatively harmless website defacements.[19] These attacks are easy to identify, and it is typically easy to recover. Others on the list are "Account Access Removal," Denial of Service, and Shutdown/Reboot. The consequences of such attacks are more severe than website defacement but less permanent than encryption and deletion. The most destructive attacks on MITRE's list are "Disk Wipe/Disk Content Wipe," "Data Destruction," and "Data encrypted for impact." Ransomware typically falls into the last two categories.

Disk Wipe/Disk Content Wipe involves wiping or corrupting information system storage to "interrupt the availability" of a system.[20] This type of attack is usually the outcome of intentionally destructive malware other than ransomware. Current

---

[17] Nelson et al. (2014).

[18] MITRE Corporation (n.d.-c) Impact.

[19] Ibid.

[20] MITRE Corporation (n.d.-b) Disk Wipe: Disk Content Wipe.

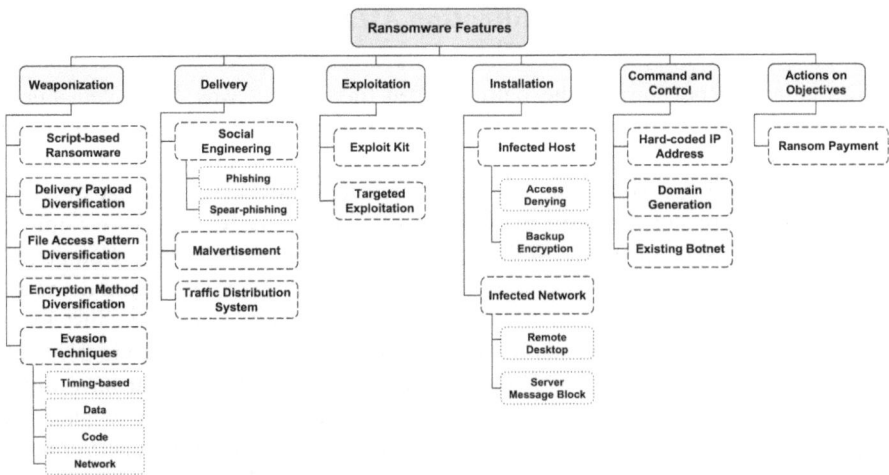

Fig. 1.1 Proposed Cyber Kill Chain-based taxonomy diagram of the ransomware features. (Daraghi et al. 2019)

ransomware business models are about profit, so attackers must convince victims that data is recoverable.

Data Destruction is the outcome of ransomware that either fails to decrypt data or intends not to decrypt data. An example of this is the XBASH attack (MITRE ID S0341).[21] Palo Alto's Unit 42 documented this attack and found "no evidence that the paid ransoms have resulted in recovery for the victims."[22] Data Destruction is different from Disk Content Wipe "because individual files are destroyed rather than sections of a storage disk or the disk's logical structure."

The MITRE ATT&CK technique that aligns most closely with ransomware is "Data Encrypted for Impact." This technique is where most ransomware falls, whether there is a recovery option or not. For example, the Shamoon 2.0 malware has two modes.[23] Mode E encrypts files, while Mode R overwrites them.[24] The intention is to destroy files in both cases.

The difference between properly functioning ransomware and other destructive attacks is that ransomware promises a path to recovery, while most other attacks in this category do not.

---

[21] MITRE Coporation (n.d.-a) Xbash.

[22] Xiao et al. (2017).

[23] Falcone (2016).

[24] Ibid.

Ransomware vs. Other Exfiltration Attacks

Exfiltration was not part of the outline when this project was conceived, but it changed the Ransomware game since the MAZE ransomware became operational around May 2019.[25] Conventional advice on recovering from a ransomware incident prior to Maze was to restore data. That all changed once MAZE began blackmailing victims. They are not the only ones doing this now. The Ransomware families WastedLocker, Netwalker, and REvil started using the same tactics. Companies with good backups and fast restoration capabilities are now paying ransom demands to avoid the public disclosure of sensitive data. Ransomware is no longer just about the threat of data loss but also about unauthorized disclosure in the public domain.

Data exfiltration is not new. Espionage and public-embarrassment have motivated hackers to exfiltrate data in the past, but extortion for financial gain has inspired a new set of criminals. Ransomware gangs steal data before encrypting it. A reasonable question is why go through the effort of exfiltrating and encrypting. The answer is that there is money to be made from both. The World Economic Forum reports that companies are 2.5 times more likely to pay a ransom in cases where data was also exfiltrated.[26] The concept of "double extortion" means that victims pay to recover their data and pay an additional amount to prevent it from leaking.[27] In one case, a company named BlackBaud stopped an attack after data exfiltration began, but before encryption took place.[28] The company paid to have the stolen data "destroyed."[29] Encryption has received the most attention, but exfiltration is a more significant threat in many cases.

Ransomware, like other attacks, uses a wide array of exfiltration techniques. The Maze ransomware by itself has at least seven:

- T1048: Exfiltration Over Alternative Protocol[30]
- T1002: Data Compressed[31]
- T1020: Automated Exfiltration[32]
- T1537: Transfer Data to Cloud Account
- T1011: Exfiltration over other network medium
- T1022: Data encrypted[33]
- T1041: Exfiltration over command and control channel

---

[25] Kennelly et al. (2020).

[26] Sayce (2024).

[27] Check Point Research (2020).

[28] Cimpanu, Cloud provider stopped ransomware attack but had to pay ransom demand anyway (2020a).

[29] Ibid.

[30] Kennelly et al. (2020).

[31] Ibid.

[32] Ibid.

[33] US Department of Health and Human Services (2020a).

Try this experiment with a crowd. Ask a group of people what is a bigger threat—hackers breaking in and stealing data or having ransomware installed on computers? This a trick question. As of 2021, they are virtually the same threat.

Technology of Ransomware

The technology that ransomware uses is the technology people commonly use every day. This section explores some of the technology and tools used in ransomware attacks and shows that attackers mostly use publicly available tools and well-known techniques for large profits.

The first ransomware broken down is LockBit. McAfee published a blog post on the software and techniques used by LockBit and deserves all of the credit for researching it. The Lockbit breakdown is summarized based on McAfee's research. https://www.mcafee.com/blogs/other-blogs/mcafee-labs/tales-from-the-trenches-a-LockBit-ransomware-story

Here are some of the key points[34]:

- Brute forced the admin password on an "outdated VPN."
- Used the SMB protocol for reconnaissance
- Used the brute-forced admin password to move laterally
- Installed ransomware on one machine and then spread it via Powershell
- Sent ARP Requests to find other hosts. Opens SMB connection to deploy ransomware. If successful, runs PowerShell to download PNG file again.
- Downloaded malicious files from the internet
- Used a .net launcher
- UAC Bypass
- Loaded modules dynamically for "detection evasion."
- Accessed the Service Manager and created its own thread to terminate processes, delete shadow volumes, and empty the recycle bin
- Attempted to stop anti-virus and databases
- Disabled boot protections
- Ransomware deletes itself using ShellExecuteExW. ShellExecuteExW is an automated way of launching applications without knowing their exact location on a computer.[35]
- Registry keys identifying victims created
- Wallpaper change
- Uses registry key and COM task schedule for persistence

---

[34] Rivero Lopez (2020).
[35] Gupta (2017).

A similar examination of the Maze Ransomware by FireEye was performed.[36] The Maze breakdown is summarized based on FireEye's research. https://www.fireeye.com/blog/threat-research/2020/05/tactics-techniques-procedures-associated-with-maze-ransomware-incidents.html

- Phishing and exploitation kits are used to get in. Phishing themes are Tax notice, Invoice, and Package Delivery.
- Phishing emails contain either malicious attachments or links to malicious documents that download Maze using macros.
- Maze also uses RDP account compromise, misconfigured internet-facing systems, and Citrix systems with weak passwords. In the Citrix case, they launched Meterpreter on the system they accessed. Meterpreter is a shell program that allows users to input commands.[37]
- Installed Cobalt Strikes "Beacon" application for Command and Control (C2). Attackers use Beacon to communicate with their control system using ordinary protocols such as HTTP, HTTPS, and DNS traffic.[38]
- Added account to the Domain Administrators group
- Used Mimikatz to "collect credentials."
- Used Bloodhound to map out the Active Directory environment and find a path to administrator access.
- Searched for files called "password" and looked for KeePass password vaults.
- Used Cobalt Strike's reconnaissance scripts
- Used batch files containing nslookup commands to map DNS
- Exfiltrated reconnaissance data in .7z files by using FTP via a PowerShell script
- Used Beacon and RDP for lateral movement
- Used a batch file to invoke taskkill command to stop anti-virus
- Used .7zip to archive files
- Began data exfiltration
- WinSCP used to exfiltrate data in some cases.
- In at least one case, replicated files to a cloud-based file systems.

The breakdowns of LockBit and Maze demonstrate two things. Attackers are generally not using unknown Zero Days. They are using tools, techniques, and procedures (TTPs) that are well known, documented, and understood. Nevertheless, they can string the various techniques together to launch effective attacks. The best explanation for how attacks such as this work came from the US Department of Health and Human Services. HHS says this about the TrickBot Trojan: "Nothing TrickBot does is unique, but its aggregate capabilities and modular flexibility make it a powerful tool."[39]

---

[36] Kennelly et al. (2020).

[37] SentinelOne (2018).

[38] Cobalt Strike by HelpSystems (n.d.).

[39] US Department of Health and Human Services (2020b).

The two situations above were avoidable, but it became common to find RDP and Citrix servers directly accessible from the internet after COVID-19 struck. It also shows that defenders who focus on doing the basics right can prevent attacks. Nothing will stop the most committed attackers, but removing sensitive systems from direct internet access, patching, and MFA is enough to avoid being an easy target. There is more involved in running an effective security program (training, monitoring, proper staffing, detection and response, and others), but an excellent place to start is with the basics.

Ransomware is not a Windows-only problem. A full list of resources showing how ransomware uses Mac OS, Linux, Android, firewalls, .Net, ICS systems, blockchain, and other technologies is provided in Appendix B: Technology of Ransomware.

Zero-Day Ransomware

Ransomware operators use an "any of the above" model for infecting victims. RDP, Citrix, VPN, and phishing are all common avenues of attack. The goal is to gain a foothold on the network, pivot, and profit. Threat actors also frequently use penetration testing tools such as Cobalt Strike and Metasploit in attacks. Both tools are publicly available and well known. There is a class of vulnerabilities exploited in attacks called "zero days." Zero Day vulnerabilities are "unknown to the computer user and software manufacturer. Since the vulnerability was previously unknown, software developers have 'zero days' to rectify the security flaw."[40]

The use of zero-days is typically reserved for nation-states and other major threat actors (APT groups, ransomware gangs, spies, and other well-financed attackers). However, they do appear in Ransomware attacks, particularly for "big game" targets. The reason for this exclusivity is the high-cost of zero-day vulnerabilities. They sell for millions of dollars and are at high risk of being discovered when used, leading to vendor patches and other mitigations. This does not mean that anyone should feel safe. The Sodinokibi gang made a splash when it first started operations by using multiple zero days in attacks.[41] Additionally, opportunistic attackers will use whatever they can get their hands on, such as an SQL vulnerability (CVE-2020-12271) found in Sophos firewalls.[42]

What can organizations do to defend themselves against zero-day vulnerabilities? The reason why zero-days are so valuable is that they are difficult to stop, but organizations can decrease their effectiveness. ENISA recommends anomaly monitoring,

---

[40] Golabek-Goldman (2014).

[41] Ashford (2019).

[42] Cimpanu, Hackers tried (and failed) to install ransomware using a zero-day in Sophos firewalls (2020c).

network segmentation, and "mitigation software."[43] All of these methods have the potential to reduce, delay, or eliminate the threat. These are all elements of "Defense-in-depth." Southern Methodist University (SMU) researchers developed a form of anomaly detection that "spots zero-day ransomware more than 95 percent of the time."[44]

> When attackers encrypt files, certain circuits inside the computer have specific types of power surges as files are scrambled. Computer sensors that measure temperature, power consumption, voltage levels, and other characteristics can detect these specific types of surges, SMU researchers found.[45]

This type of tool would be valuable to most organizations if it could stay ahead of the cat and mouse game with attackers continually adjusting their techniques. The patent application number is USPTO 15/812663.[46] The zero-day topic goes back to the broader malware topic. If attackers can use zero-day vulnerabilities to install ransomware, they can use them for almost any other kind of attack. Organizations must defend against zero-day attacks, one of which results in ransomware.

---

[43] ENISA (n.d.).

[44] Taylor et al. (2020).

[45] Ibid.

[46] Thornton et al. (2020).

Chapter 2
The Three Cs of Ransomware

Contents

Abstract In this chapter, we introduce the "Three Cs of Ransomware," a novel conceptual framework designed to unpack the potency and profitability of ransomware attacks. By examining Cryptography, Cryptocurrency, and Customer Service, we reveal how these components collectively contribute to the success of ransomware operations. Cryptography ensures data is encrypted and inaccessible, and thus central to the extortion process, while Cryptocurrency enables anonymous financial transactions that are difficult to trace, facilitating the payment of ransoms. The aspect of Customer Service highlights the evolution of ransomware into a sophisticated business model that includes ransomware as a Service (RaaS), enhancing the efficiency of these criminal enterprises. This framework sheds light on the mechanisms that make ransomware a formidable cybersecurity challenge and emphasizes the interconnectedness of these elements in ensuring the success and profitability of ransomware attacks. Through this analysis, the chapter aims to provide a deeper understanding of the strategic operation of ransomware and the critical areas for defensive strategies. This chapter aims to show how ransomware gangs use cryptography and cryptocurrency with a touch of customer service.

Three Cs: Cryptography

The first of the Three C's of Ransomware covered is Cryptography. Ross Anderson defines cryptography as the "science and art of designing ciphers" in his text called "Security Engineering."[1] It is the technique that makes ransomware profitable. Encryption enciphers data so that it is not readable without decryption. A ransomware attack without encryption would fail in its goal of extracting money. There are many failed ransomware attacks, but one of the most famous is WannaCry. WannaCry spread rapidly through networks in 2017, encrypting computers until a "kill switch" was found after a few hours.[2] The kill switch blocked the encryption process. WannaCry continued to spread as a lame worm, becoming the "largest global ransomware attack in internet history" at the time.[3] However, it failed to achieve its revenue potential because of the rapid discovery of the kill switch. It collected 54.43228033 BTC equaling approximately \$386,905 as of December 2, 2019.[4] This is a significant revenue, but not compared to \$61 million for Ryuk, or even \$8 million for Bitpaymer as of 2020.[5] Ransomware without an encryption (or extortion) scheme is just another piece of malware (Fig. 2.1).

Ransomware can use both symmetric and asymmetric encryption, often both in the same attack. Symmetric encryption uses a single key for encryption and decryption. This makes it difficult for ransomware authors to protect the key. Ransomware that only uses symmetric encryption must store the key on a computer in a way that avoids discovery by investigators that would subsequently transfer it back to the threat actor. One of the most egregious cases of mishandling symmetric keys is the "EffectiveIdiot" mac ransomware. SentinelOne researchers stated, "The clear text key used for encoding the file encryption key ends up being appended to the encoded file encryption key. Taking a look at a completely encrypted file shows that a block of data has been appended to it."[6] Another example is the Harasom ransomware, which hid the symmetric key in the ransomware executable, making it "easy" to discover.[7] This is similar to how researchers broke DRM for HD-DVD/Blu-ray, where they found encryption keys in computer memory during playback.[8]

The flaws do not mean that symmetric encryption does not have a role, nor does it mean that asymmetric encryption is flawless. Asymmetric encryption uses a pair of keys known as public and private keys. Operators can freely share public keys, while maintaining the secrecy of private keys. Ransomware can use asymmetric encryption alone, but it is more common to use a technique similar to the TLS

---

[1] Anderson (2008).

[2] UK Natoinal Audit Office Comptroller and Auditor General (2018).

[3] Brandom (2017).

[4] Web Titan (2019).

[5] Spadafora (2020).

[6] SentinelOne (2020).

[7] Sarah (2017).

[8] Beschizza (2007).

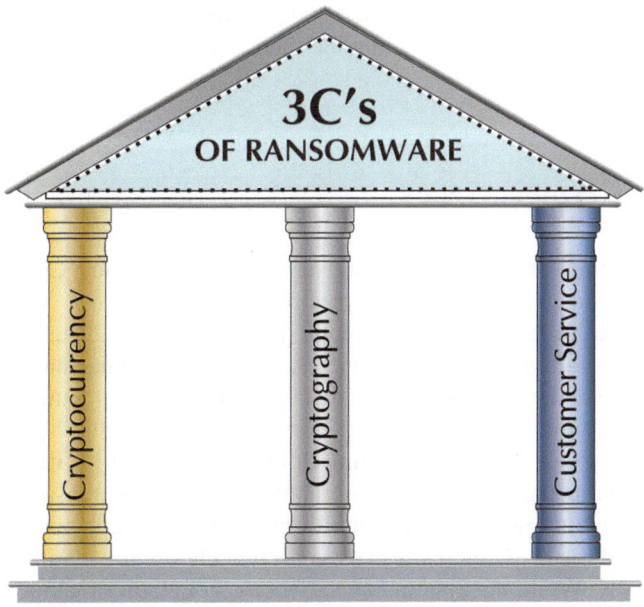

Fig. 2.1 Graphic showing the Three C's of Ransomware as pillars

handshake.[9, 10] In TLS, asymmetric encryption protects a "handshake" process that generates a symmetric key agreed upon between the client and the server. The most successful ransomware uses a similar process.

Sodinokibi/REVIL handles encryption effectively, as detailed by Acronis. Sodinokibi/REVIL use two public keys.[11] It uses the Elliptic-curve Diffie-Hellman (ECDH) algorithm with each public key to create session keys and then different AES and Salsa20 symmetric keys.[12] Sodinokibi uses the AES key to protect the private keys and the Salsa20 key. The Salsa20 key encrypts victims' files.[13]

Flaws in encryption are an avenue to explore, but they should not be the core part of an incident response plan. Ransomware occasionally uses custom encryption code.[14] More recent and sophisticated ransomware uses "existing libraries."[15] During an investigation, malware analysis might uncover the type of encryption used, leading to the discovery of a flaw or decryption key. Leading ransomware investigator Fabian Wosar says that his company [Emsisoft] tries to find

---

[9] Puodzius (2016).

[10] Cloudflare (n.d.).

[11] Tiwari (2020).

[12] Bernstein (2008).

[13] Tiwari (2020).

[14] Melendez (2016).

[15] Ibid.

cryptographic flaws in new ransomware and generates free decryption tools if they find any.[16] However, "so you're telling me there's a chance" is no substitute for an incident response plan.[17]

Three Cs: Cryptocurrency

The second of the Three C's of Ransomware is Cryptocurrency. Cryptocurrency is the high-level name for a category of electronic cash. The word "cryptocurrency" derives from the combination of technology (cryptography) and purpose (currency). There are many different cryptocurrencies. The most popular and capitalized is Bitcoin.[18] Bitcoin was founded by "Satoshi Nakamoto" as a "purely peer-to-peer version of electronic cash [that] would allow online payments to be sent directly from one party to another without going through a financial institution."[19] Bitcoin uses blockchain technology that cryptographically links one transaction to another in a chain.[20] Bitcoin transactions are public, but they do not contain any identifying information unless the participants in a transaction choose to identify themselves in some other way. While the system is anonymous by default, tracking technologies can still link transactions to individuals' real identities.[21] For example, the cybersecurity firm Hyas was able to identify "well-known brokers that make payments on behalf of victims."[22] The following section examines the use of cryptocurrencies to facilitate ransomware payment, including steps criminals take to evade identification and preparations companies should make to recover from an incident.

Ransomware works by quietly encrypting files on a computer and then displaying a ransom note that demands payment using cryptocurrency, often within a set time. Figure 2.2 below is a Sodinokibi ransom note. The note explains what happened, the ransom amount, bitcoin address to send funds, and bitcoin purchasing instructions. It also contains a threat to double the ransom if it is not paid within two days.

Ransom notes all follow a similar theme. Some contain all of the information required to resolve the incident, while others point to contact information such as email addresses, chat services, and websites required to pay the ransom or begin the ransom payment process.

There are many cryptocurrencies, but Bitcoin dominates in regard to ransom payments. A report from the blockchain investigative firm Ciphertrace found that

---

[16] Johnson (2019).

[17] Carrey (1994).

[18] Ang (2021).

[19] Nakamoto (2008).

[20] Meiklejohn et al. (2013).

[21] LR and AO (2018).

[22] HYAS Intel Team (2021).

Fig. 2.2 Sodinokibi ransom note breakdown

bitcoin was used in 98% of ransomware attacks.[23] One percent of the attacks used another cryptocurrency called Monero, while the rest used a collection of others.

Monero is one of a class of "privacy coins" that also includes Zcash and Dash.[24] The benefit of "privacy coins" is that transaction attributes are concealed, making the participants harder to trace and identify because of "mixing and cryptographic enhancements."[25, 26] Mixing is an online money laundering technique. Owners transfer cryptocurrencies into a mixer from a traceable point. The currencies become difficult to trace because mixing services split them into smaller quantities and transfer them into random wallets.

The US agency FinCEN refers to cryptocurrencies as "Convertible Virtual Currencies" (CVCs) with a subcategory of privacy coins known as

---

[23] CipherTrace Cryptocurrency Intelligence (2019).

[24] US Department of Justice (2020a, b, c).

[25] PYMNTS (2018).

[26] FinCEN (2020).

Fig. 2.3 Cryptocurrency
used to pay ransoms by
percentage. (Based on:
https://ciphertrace.com/
wp-content/
uploads/2019/09/
CipherTrace-
Cryptocurrency-Anti-
Money-Laundering-
Report-2019-Q2-3.pdf)

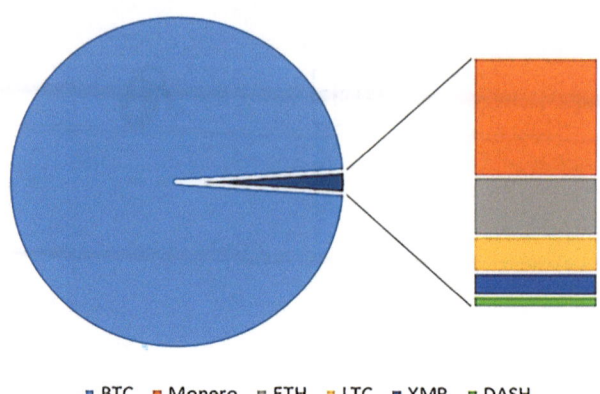

Ransomware Cryptocurrency by percentage

■ BTC   ■ Monero   ■ ETH   ■ LTC   ■ XMR   ■ DASH

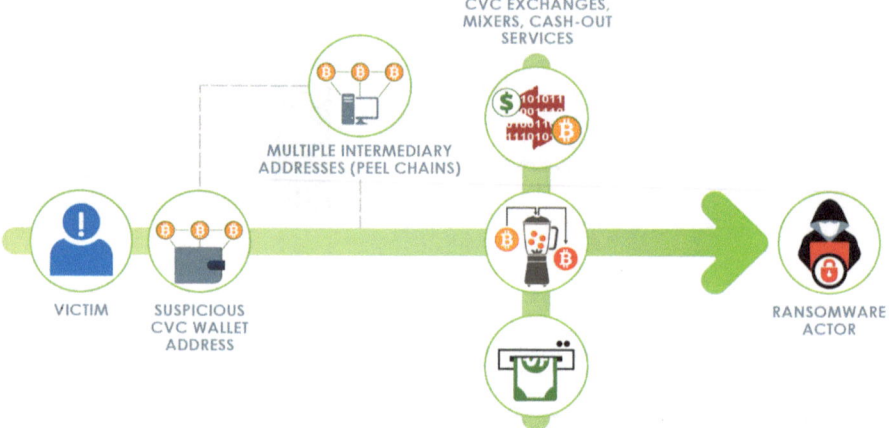

Fig. 2.4 "Movement of CVCs in ransomware attacks". (Source: https://www.fincen.gov/sites/
default/files/advisory/2020-10-01/Advisory%20Ransomware%20FINAL%20508.pdf)

"Anonymity-Enhanced Cryptocurrencies" (AECs).[27] FinCEN found that ransomware operators sometimes provide discounts to victims paying in AECs.[28] They also
convert Bitcoin into privacy coins using an "exchange" service, making tracking
more difficult.[29] Figure 2.3 shows how cryptocurrencies flow from victims through
exchanges, mixers, and cash-out services to ransomware "actors" (Fig. 2.4).

Many different groups are trying to get their piece of the profits. Ransomware
operators develop and maintain the ransomware. Ransomware affiliates infect

---

[27] Ibid.

[28] Ibid.

[29] Ciphertrace (2021).

victims and collect most of the profits. They also sometimes hire specialized hacking groups known as Initial Access Brokers (IABs) to breach networks. Mixers and exchanges collect service fees. Even insurance companies that make payments are making a profit. The insurance company AON reported that cyber insurance policies are more profitable than property and casualty insurance policies.[30] Ciphertrace charted how money flows in the Netwalker ransomware attack (Fig. 2.5):

This high-level diagram shows how bitcoin moves from a victim (or their representative) into a bitcoin wallet of the attacker.[31] The "Ransomware as a Service" affiliate and its operators split up the profits. Notice that those causing the infection receive 80% while the ransomware developers/service providers get 20%.

Chainalysis tracked the same activity with Sodinokibi (Fig. 2.6):

Chainalysis also tracked the way ransoms were "cashed out." The details on how cashing out works are beyond the scope of this paper.

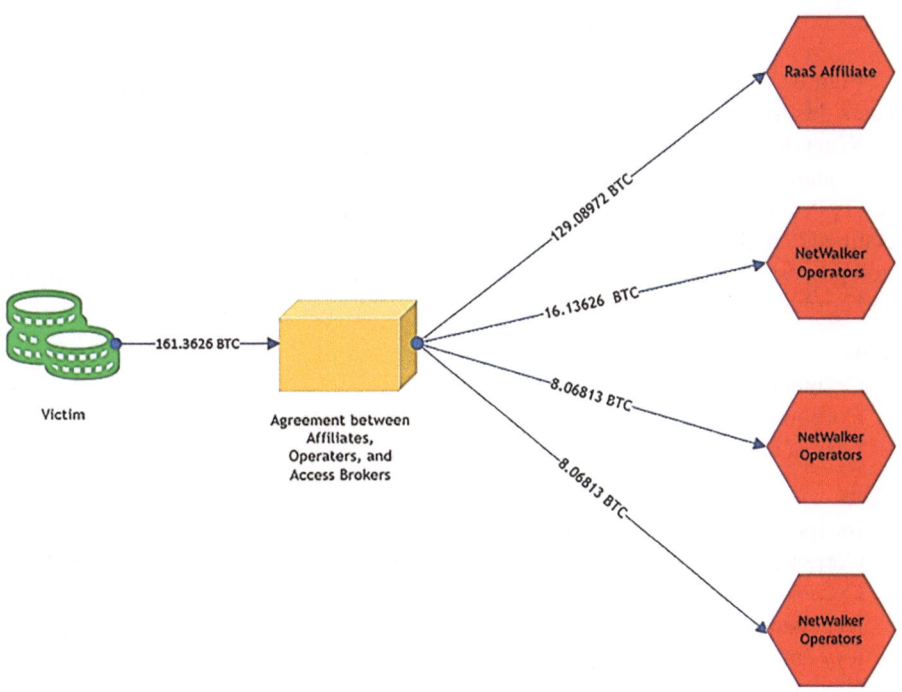

Fig. 2.5 "Example of ransom payments split between RaaS operators and the affiliate who caused the infection." (Based on: Ciphertrace Cryptocurrency Intelligence https://ciphertrace.com/tracing-ransomware-ciphertrace-helps-mcafee-follow-netwalker-funds/)

---

[30] Dudley, The Extortion Economy: How Insurance Companies Are Fueling a Rise in Ransomware Attacks (2019a).

[31] Pamela (2020) Ciphertrace.

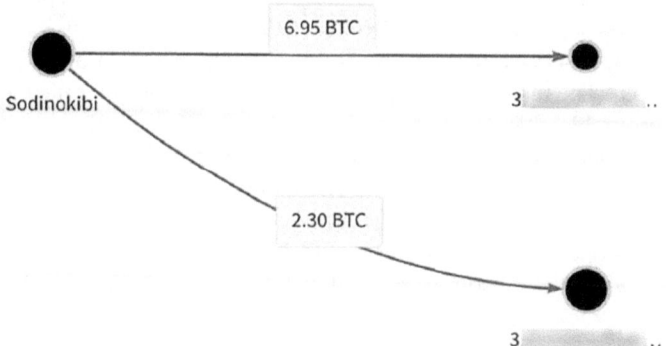

Fig. 2.6 "Sodinokibi RaaS user on the left sends 70–75% of ransoms taken to one address, likely their own, and 20–25% to another address, likely controlled by the RaaS vendor." (Source: https://go.chainalysis.com/2020-Crypto-Crime-Report.html)

Figure 2.6 shows significant activity on Russian exchanges. Chainalysis attributes this to the Cryptolocker, Locky, and Cerber strains.[32]

What does this all mean for businesses and victims?

Organizations should prepare for the possibility of paying ransoms. Should they stockpile bitcoin as 60% of CEOs in the 2018 Code 42 Data Exposure report claimed?[33] This is one of the most important questions in the ransomware preparation process. It requires collaboration between Cybersecurity, executive leadership, and corporate finance (Fig. 2.7).

Pros
- Cryptocurrency on hand if needed
- Corporate finance is familiar with purchasing cryptocurrency
- Less of a scramble if the need to use cryptocurrency arises

Cons
- Inexperience with negotiation with threat actors can worsen an incident
- Corporations should not make ransom payments without external legal advice
- Corporations probably do not want ransom payments directly attributable to them
- Regulatory risk: Sanctions compliance, suspicious activity exposure, and regulatory enforcement actions
- Negative publicity if payment is publicized

---

[32] Chainalysis (2020).
[33] Code (422018).

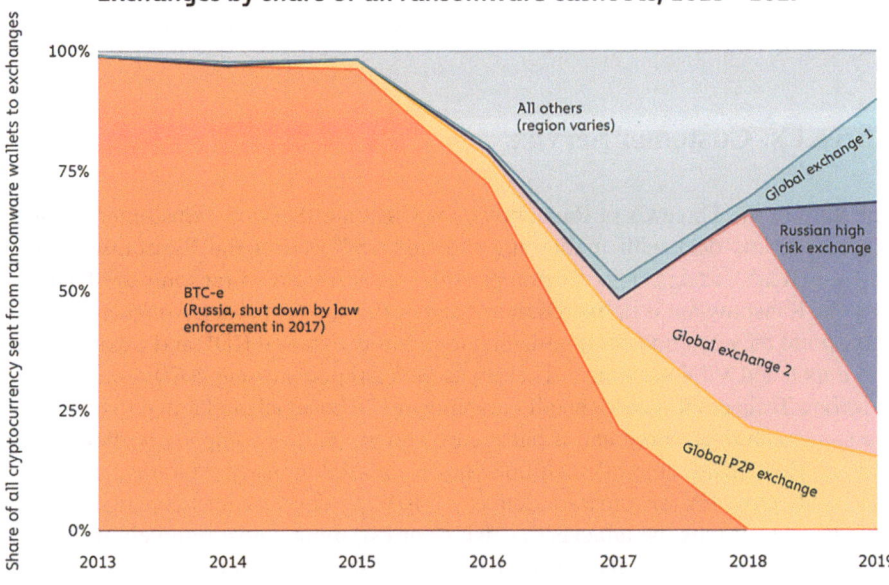

Fig. 2.7 Share of all cryptocurrency sent from ransomware wallets to exchanges. Source: https://go.chainalysis.com/2020-Crypto-Crime-Report.html

Cryptocurrency Recommendations

Organizations should be knowledgeable about cryptocurrencies and have access to them if necessary. The amounts can be small. The point is setting up the policies and processes for managing cryptocurrencies in case the need arises. Forrester Research recommends having a plan for acquiring and paying bitcoin.[34]

Organizations should consider whether they will make direct payments to criminals or use a third party. Cyber insurance and specialized incident response firms can assist with facilitating payments through authorized third parties. Organizations should be aware of third-party negotiators approved by their cyber-liability insurance policy and include them in their incident response plan. Both services can be "life-savers" in the heat of an incident.[35] However, organizations should confirm and obtain approval from their insurance carriers before making payments or hiring a third party firm to assist with negotiation and payment.[36]

Criminals use ransomware because they want to make money. Paying a ransom is one of the most uncomfortable activities a company will ever face. However, it is an increasing probability with the growing rate of ransomware infection and

[34] Zelonis (2019).

[35] LaCroix (2019).

[36] Ibid.

increasing profitability for criminals.[37] Organizations must prepare for the likeli-
hood of facing a ransom or extortion demand and the possibility of paying.

Three Cs: Customer Service

The third of the Three Cs of Ransomware is Customer Service. "Customer service"
is who victims deal with if they are infected with one of the Ransomware as a
Service (RaaS) strains. This section provides an overview of ransomware business
models. Ransomware is delivered in an "any of the above" model. Attackers use any
entry point they can find, from phishing to publicly exposed RDP and Citrix servers
and unpatched VPN systems.[38] There was also a report in June 2020 of an attempt
to bribe a Tesla employee to install ransomware.[39] The attacker claimed to have used
the same technique three and a half years ago at another company without being
caught. This "insider threat" technique has been used in many espionage cases and
can be effective for ransomware delivery.[40] Bribing employees is considered "bra-
zen" because it puts the attackers at risk of arrest, unlike most remotely run large-
scale ransomware incidents.[41] The image below shows someone offering to install
ransomware on their network for 3 bitcoin (Fig. 2.8).[42]

Ransomware operations have become more business-like and sophisticated as
they have grown from a model of individual computer compromise by one or more
attackers to "affiliate" operations in an "as a service" (RaaS) service model.
Ransomware gangs experienced "skyrocketing profits" due to the adoption of RaaS
delivery models.[43] The RaaS business model resembles other types of online businesses. Ransomware operators are comparable to cloud service providers. They create the ransomware and handle payment, and support. They also handle recruitment
with robust marketing operations. Ransomware "affiliates" are similar to partners or
integrators. They identify and compromise victims. The financial incentives are
high for both sides, with operators taking a 20% share of the total revenue while
affiliates get 80%.

Ransomware gang REvil/Sodinokibi made a marketing splash by depositing 99
bitcoins (worth approximately one million dollars at the time) into a Russian language hacker forum in September 2020.[44] The deposit is a show of strength and

---

[37] Coveware (2020).

[38] Cimpanu, Top exploits used by ransomware gangs are VPN bugs, but RDP still reigns supreme
(2020e).

[39] Greenberg (2020).

[40] FBI (2015).

[41] Greenberg (2020).

[42] Hanslovan (2020).

[43] Intel 471 (2020).

[44] Thompson (2020).

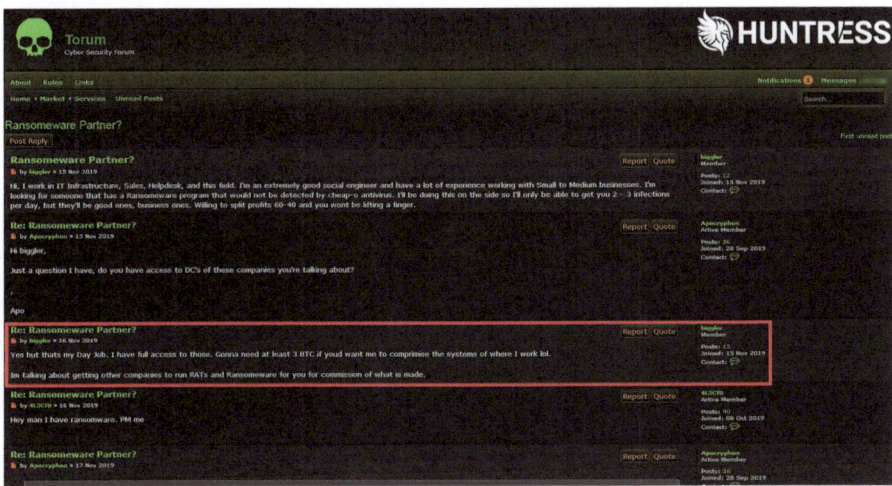

Fig. 2.8 Forum post offering remote access to victims for 3BTC. (Source: Hanslovan 2020)

evidence that affiliates can succeed. Meanwhile, the Lockbit gang took a different approach in June 2020. They launched a paper writing competition with a fifteen thousand dollar payout. The competition is beneficial because ransomware operators learn the latest hacking techniques while building their networks and flaunting their money. Furthermore, they demonstrate the sophistication of their operation, which makes it a powerful recruitment technique (Fig. 2.9).

The next image shows the marketing site for Ranion. FUD stands for "fully undetectable" and is a widely used term in malware marketing (Fig. 2.10).

There are two "package options" (Fig. 2.11):

The more expensive package buys more time for a small savings. This demonstrates Ranion gangs' expectation of longer-term operations.

The following image of the Cryptonite ransomware shows that it has three packages, each with more robust features and the ability to collect higher ransoms (Fig. 2.12).

The next image shows how the "subscription" model works for Ghostly Locker ransomware (Fig. 2.13).

The images demonstrate that victims face organized and well-prepared adversaries in many cases.

The latest innovation in ransomware operations is the use of specialized "Network Access Sellers," also known as "Initial Access Brokers." According to Accenture, "Network Access Sellers' expertise lies in the ability to gain corporate and government network access, which they then sell to other cybercrime groups for a handsome profit."[45] Profiles of compromised networks are posted on Dark Web forums for $300–$10,000. Accenture says that they see RDP, Citrix, and VPN compromises

---

[45] Accenture (2020).

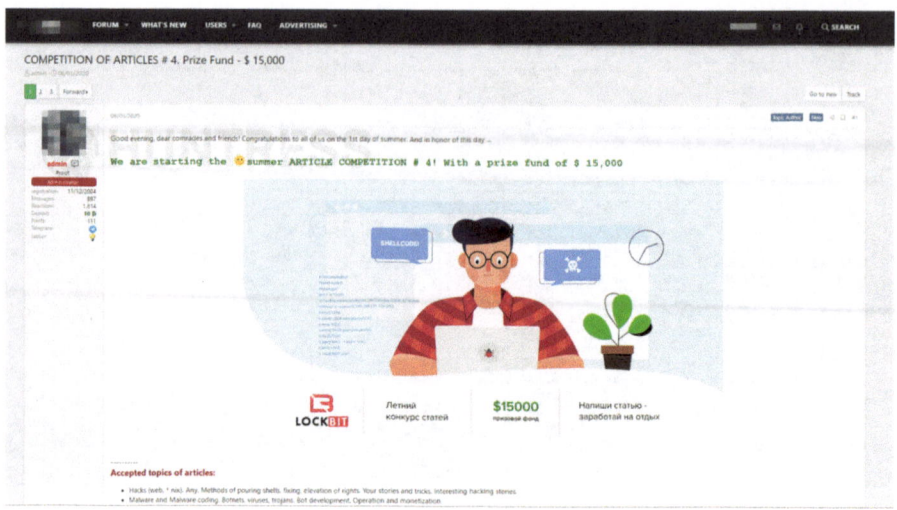

Fig. 2.9 Paper-writing competition sponsored by the LockBit ransomware gang. (Source: Hanslovan 2020)

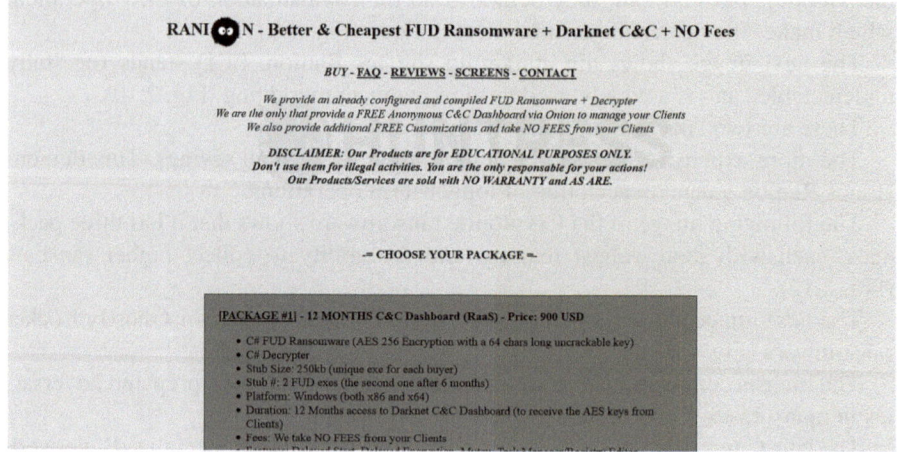

Fig. 2.10 Marketing site for Ransion ransomware offering "FUD: Fully Undetectable" Ransomware. (Source: Hanslovan 2020)

[PACKAGE #2] - 6 MONTHS C&C Dashboard (RaaS) - Price: 490 USD

- C# FUD Ransomware (AES 256 Encryption with a 64 chars long uncrackable key)
- C# Decrypter
- Stub Size: 250kb (unique exe for each buyer)
- Stub #: 1 FUD exe
- Platform: Windows (both x86 and x64)
- Duration: 6 Months access to Darknet C&C Dashboard (to receive the AES keys from Clients)
- Fees: We take NO FEES from your Clients
- Features: Delayed Start, Delayed Encryption, Mutex, Task Manager/Registry Editor Disabler, UAC Bypass, Desktop Wallpaper Changer
- IP Tracking: Yes
- Offline Encryption: Yes
- Support: Yes
- Real-Time Client Manager: Yes
- Paid Add-On (Dropper): Execute your own exe (backdoor, implant, etc.) (+90 USD)
- Paid Add-On (Clone): A fresh FUD RANION copy with the same setup information (+90 USD)
- Paid Add-On (Crypter): Additional Crypter/Obfuscator + unique onion address (+90 USD)
- Paid Add-On (Unkillable Process): Unkillable Process aka BSOD (+90 USD)
- Free Add-On: optional file types to encrypt (for all encrypted file types see FAQ)
- Free Add-On: optional Client's sub-banner in your language (already present en, ru, de, fr, es, it, nl, fas, za)

[PACKAGE #1] - 12 MONTHS C&C Dashboard (RaaS) - Price: 900 USD

- C# FUD Ransomware (AES 256 Encryption with a 64 chars long uncrackable key)
- C# Decrypter
- Stub Size: 250kb (unique exe for each buyer)
- Stub #: 2 FUD exes (the second one after 6 months)
- Platform: Windows (both x86 and x64)
- Duration: 12 Months access to Darknet C&C Dashboard (to receive the AES keys from Clients)
- Fees: We take NO FEES from your Clients
- Features: Delayed Start, Delayed Encryption, Mutex, Task Manager/Registry Editor Disabler, UAC Bypass, Desktop Wallpaper Changer
- IP Tracking: Yes
- Offline Encryption: Yes
- Support: Yes
- Real-Time Client Manager: Yes
- Paid Add-On (Dropper): Execute your own exe (backdoor, implant, etc.) (FREE)
- Paid Add-On (Clone): A fresh FUD RANION copy with the same setup information (+90 USD)
- Paid Add-On (Crypter): Additional Crypter/Obfuscator + unique onion address (+90 USD)
- Paid Add-On (Unkillable Process): Unkillable Process aka BSOD (+90 USD)
- Free Add-On: optional file types to encrypt (for all encrypted file types see FAQ)
- Free Add-On: optional Client's sub-banner in your language (already present en, ru, de, fr, es, it, nl, fas, za)

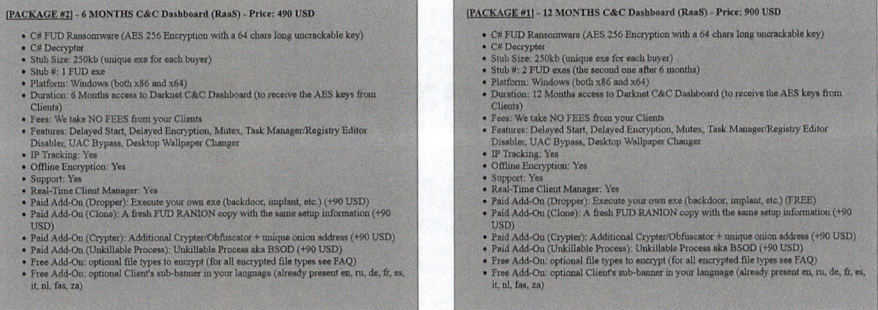

Fig. 2.11 Ranion ransomware packaged offerings. (Source: Hanslovan 2020)

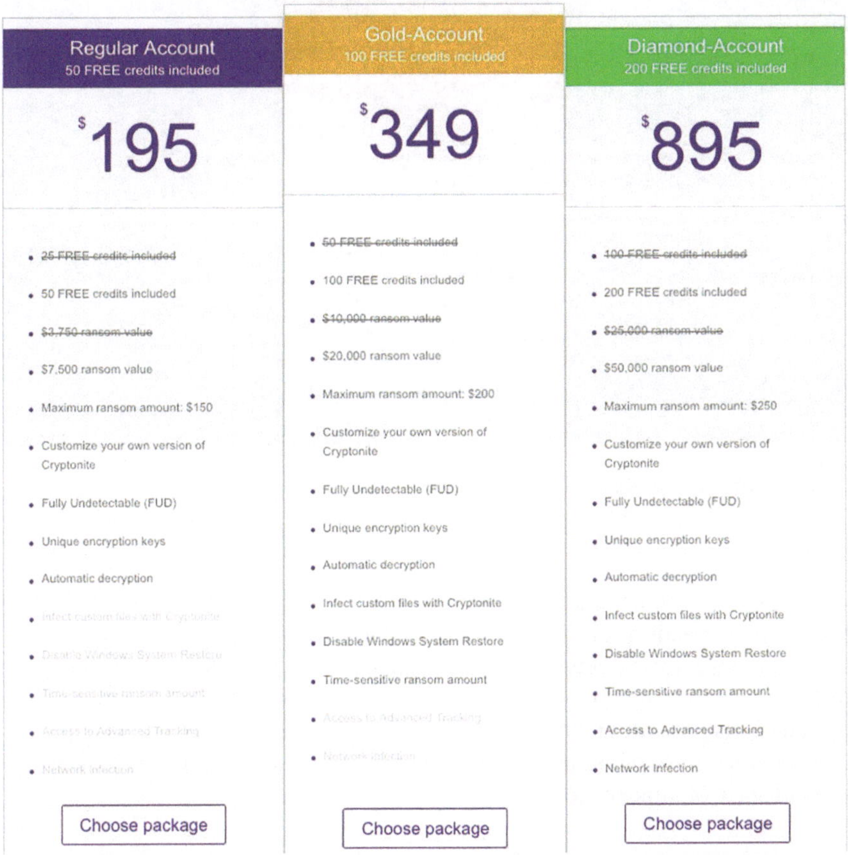

Regular Account
50 FREE credits included
$195
25 FREE credits included
50 FREE credits included
$3,750 ransom value
$7,500 ransom value
Maximum ransom amount: $150
Customize your own version of Cryptonite
Fully Undetectable (FUD)
Unique encryption keys
Automatic decryption
Infect custom files with Cryptonite
Disable Windows System Restore
Time-sensitive ransom amount
Access to Advanced Tracking
Network infection
Choose package
Gold-Account
100 FREE credits included
$349
50 FREE credits included
100 FREE credits included
$10,000 ransom value
$20,000 ransom value
Maximum ransom amount: $200
Customize your own version of Cryptonite
Fully Undetectable (FUD)
Unique encryption keys
Automatic decryption
Infect custom files with Cryptonite
Disable Windows System Restore
Time-sensitive ransom amount
Access to Advanced Tracking
Network infection
Choose package
Diamond-Account
200 FREE credits included
$895
100 FREE credits included
200 FREE credits included
$25,000 ransom value
$50,000 ransom value
Maximum ransom amount: $250
Customize your own version of Cryptonite
Fully Undetectable (FUD)
Unique encryption keys
Automatic decryption
Infect custom files with Cryptonite
Disable Windows System Restore
Time-sensitive ransom amount
Access to Advanced Tracking
Network Infection
Choose package

Fig. 2.12 Cryptonite RaaS packages

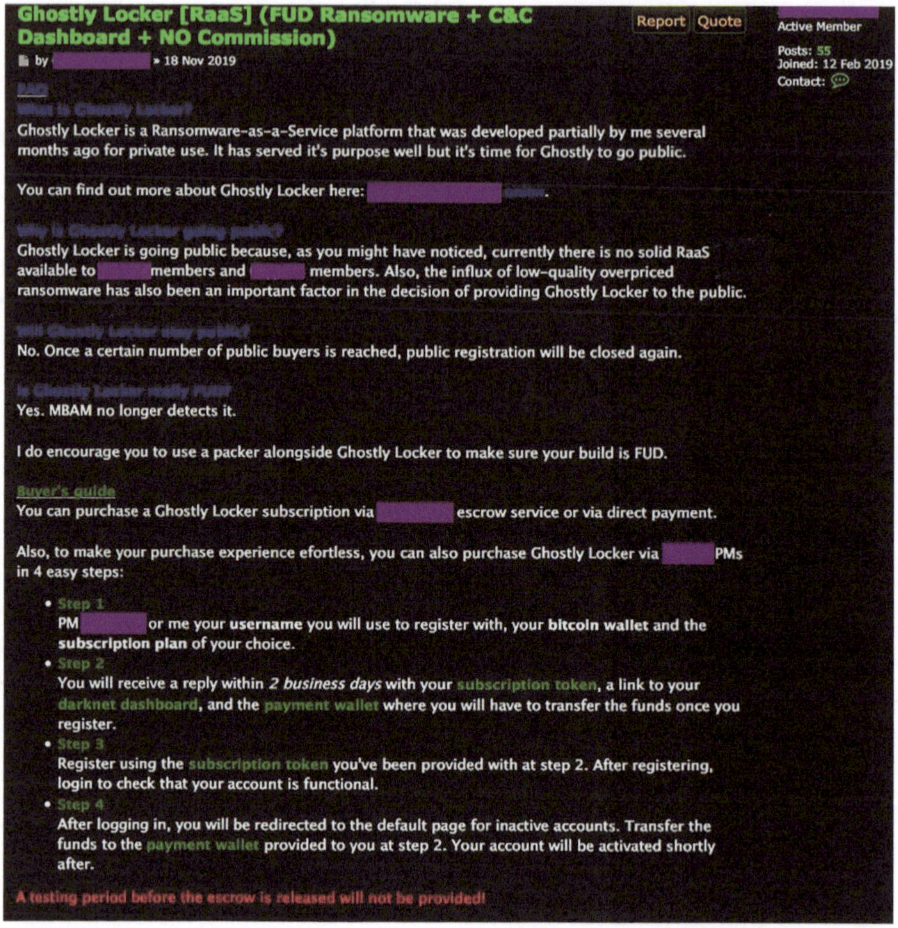

Ghostly Locker [RaaS] (FUD Ransomware + C&C
Dashboard + NO Commission)

Report   Quote

Active Member

📄 by   » 18 Nov 2019

Posts: 55
Joined: 12 Feb 2019
Contact: 💬

RaaS

What is Ghostly Locker?

Ghostly Locker is a Ransomware-as-a-Service platform that was developed partially by me several months ago for private use. It has served it's purpose well but it's time for Ghostly to go public.

You can find out more about Ghostly Locker here: ▮▮▮▮▮▮▮ ▮▮▮▮.

Why is Ghostly Locker going public?

Ghostly Locker is going public because, as you might have noticed, currently there is no solid RaaS available to ▮▮▮▮ members and ▮▮▮▮ members. Also, the influx of low-quality overpriced ransomware has also been an important factor in the decision of providing Ghostly Locker to the public.

Will Ghostly Locker stay public?

No. Once a certain number of public buyers is reached, public registration will be closed again.

Is Ghostly Locker really FUD?

Yes. MBAM no longer detects it.

I do encourage you to use a packer alongside Ghostly Locker to make sure your build is FUD.

Buyer's guide

You can purchase a Ghostly Locker subscription via ▮▮▮▮ escrow service or via direct payment.

Also, to make your purchase experience effortless, you can also purchase Ghostly Locker via ▮▮▮ PMs in 4 easy steps:

- **Step 1**
  PM ▮▮▮ or me your **username** you will use to register with, your **bitcoin wallet** and the **subscription plan** of your choice.
- **Step 2**
  You will receive a reply within *2 business days* with your subscription token, a link to your darknet dashboard, and the payment wallet where you will have to transfer the funds once you register.
- **Step 3**
  Register using the subscription token you've been provided with at step 2. After registering, login to check that your account is functional.
- **Step 4**
  After logging in, you will be redirected to the default page for inactive accounts. Transfer the funds to the payment wallet provided to you at step 2. Your account will be activated shortly after.

A testing period before the escrow is released will not be provided!

Fig. 2.13 Ghostly Locker RaaS subscription model

advertised and are seeing attackers "increasingly diversifying their methods," which "substantially expands a corporation's exploitable attack surface."[46] Network Access Sellers sell compromised networks to the highest bidder regardless of the attacker's motivations. Ransomware has injected a new class of well-funded purchasers into the system. This reiterates that network defenders should prioritize malware prevention and defense-in-depth as the key goals because a breached network leaves itself open to all sorts of attacks, not just ransomware.

---

[46] Idib.

Chapter 3
Business Impacts of Ransomware

Contents

Abstract This chapter comprehensively examines ransomware's the impact of ransomware on organizations, highlighting the human factor as a critical element in ransomware attacks and defending against them. Ransomware, described as a pervasive threat to society, is operated by organized groups aiming for profitability, akin to business entities, and relies heavily on human decisions at various stages, from execution by operators and affiliates to the responses of victims. Chapter 4 includes a survey conducted between October and December 2020, involving participants from diverse roles and industries to investigate attitudes toward ransomware. This study revealed significant insights, such as the prevalence of incident response plans among organizations, albeit with a notable percentage lacking specific plans for ransomware. Surprisingly, none of the surveyed organizations had reserved cryptocurrency for ransom payments, contrasting with earlier reports suggesting a trend of stockpiling cryptocurrency as a preparative measure against ransomware attacks. The findings underscore the complex interplay of preparation, decision-making, and response strategies in mitigating ransomware risks. Additionally, the study explores the varied impacts of ransomware across sectors, emphasizing the vulnerability of small to medium-sized businesses (SMBs), the education sector, and healthcare, as well as the challenges managed service providers (MSPs) face. It concludes with an overview of relevant laws and regulations, highlighting the legal landscape surrounding ransomware payments and the evolving strategies to combat this form of cybercrime.

Human Factors

Ransomware is the ultimate human factor problem. Everyone involved in a ransomware event contributes to the incident and the outcome.

- Ransomware operators organize themselves similarly to businesses for efficiency and profitability.
- Ransomware Affiliates are like plumbers and electricians. They connect victims to the service for a cut of the profits.
- Decisions made by victims over a long period make it possible for ransomware to root itself in society. Victimization is almost all about human factors. Security decisions leading up to an incident, the level of training, and the actions that enable or prevent ransomware are all consequences of human factors.
- Victims decide how they will respond. Will they pay, restore, go out of business, or a combination of all three?

A survey was conducted from October to December 2020 as part of this research to explore attitudes and myths about ransomware. Eighty-three participants in a variety of roles and industries completed the survey. The full survey is in Appendix J. A selection of survey responses are presented, with the respondents identifying their roles as follows (Tables 3.1, 3.2, and 3.3):

One of the challenges in ransomware response is the lack of incident response plans. 95.65% of respondents stated that their organization has an incident response plan (Table 3.4), while 86.21% of those plans include ransomware (Table 3.5).

The results are broken out by organization size. Ransomware is present in incident response plans for organizations of all sizes, but is far more common in large organizations (1000+ users) as shown in Table 3.6. The number of organizations that do not address ransomware is also significant because they are susceptible to the largest payouts.

The survey asked, "Does your organization have bitcoin set aside for ransomware payments? One hundred percent of respondents said no. This is in contrast with The Code42 2018 Data Exposure Report. The report found that 73% of CISOs and 60% of CEOs claimed that their company was "currently stockpiling or has in the past 12 months stockpiled cryptocurrency in case of a ransomware attack/data

Table 3.1 Survey question asking about the respondents role

Q30	In your role at your organization, are you a:		
#	Answer	%	Count
1	Business Leader (COO, CMO, VP, Director, etc.)	6.49%	5
4	CEO	2.60%	2
6	Security & IT Leader (CISO, CIO, CTO, CDO, VP, Director, etc.)	37.66%	29
7	IT/Security Staff	46.75%	36
8	Non-IT/non-Security Staff	6.49%	5
	Total	100%	77

Table 3.2 Survey question asking about the respondents organization type

Q1	Do you work for a for-profit, non-profit or government entity?		
#	Answer	%	Count
1	For-profit	78.31%	65
2	Non-profit	7.23%	6
3	Government	10.84%	9
4	Don't work/Unemployed/Retired/Etc.	3.61%	3
	Total	100%	83

Table 3.3 Survey question asking about how many computer users are within the organization

Q2	How many users are at your organization?		
#	Answer	%	Count
4	1–9	6.49%	5
5	10–49	5.19%	4
6	50–99	3.90%	3
7	100–249	2.60%	2
8	250–499	6.49%	5
9	500–999	0.00%	0
10	1000–4999	23.38%	18
11	5000–9999	6.49%	5
12	10,000+	45.45%	35
	Total	100%	77

breach."[1] The Code42 number seems high. The actual number is probably somewhere in the middle.

An essential part of ransomware incident response is having a third-party ready to handle forensics, potential negotiations, vetting of the threat actor, and facilitating payment. A cyber insurance carrier can handle this in many cases. Twenty-five percent of survey respondents claimed to have a contract with a third party for this purpose (Table 3.7).

The next two questions explored the scenarios where organizations would consider paying ransoms.

Question 11 asked whether participants' organizations would pay for any listed reason (Table 3.8). The responses show why the extortion aspect of ransomware is so powerful. The leading answer is "No, I would never pay," but closely behind that is "if there was no other choice because backups do not exist or work" or "to regain access to files that the company has no other way of accessing." The results of this question demonstrate once again how important it is to have offline, "immutable," and protected backups.[2] It would be interesting to explore whether organizations would be willing to go out of business because they are opposed in principle to

---

[1] Code 42 (2018).

[2] Abrams, Ransomware Attackers Use Your Cloud Backups Against You (2020d).

Table 3.4 Survey question about whether the respondents organization has an incident response plan

Q5	Does your organization have an incident response plan?		
#	Answer	%	Count
1	Yes	95.65%	66
2	No	4.35%	3
	Total	100%	69

Table 3.5 Survey question asking whether the respondents organization has an incident response plan addressing ransomware

Q6	Does your organization have an incident response plan for ransomware?		
#	Answer	%	Count
1	Yes	86.21%	50
2	No	13.79%	8
	Total	100%	58

paying a ransom. There is evidence of organizations going out of business because they cannot afford to pay a ransom or because it is too complicated to restore.[3]

There are several forms of pressure during a ransomware incident. One of them is regaining access to data. The other is restoring operations with minimal downtime. The survey asked, "Would your organization pay a "reasonable" ransom demand if it would take longer than a certain amount of time to recover operations? (Table 3.9)".

The time to recover question examines the ransomware problem from a different perspective. A total of 25.81% of respondents in question 11 claimed that they would never pay a ransom for any reason related to the feasibility of recovery. 55.17% would not pay a ransom to reduce the amount of time it took to recover. Incident response planners should include these two questions in tabletop exercises with decision makers. Many people say that they will never pay because they have backups. This does not reflect the realities of how complex and time consuming it is to recover, or the possibility that hackers will attack backups.

Ransomware Impact on Sectors

One of the motivations for pursuing this research was the 2019 Summer of Ransomware attacks against schools and its damaging and often fatal impacts on small businesses. The diversity of victims grew over time to encompass organizations from the corporate world to hospitals, Managed Service Providers (MSPs),

---

[3] Muncaster (2020).

Table 3.6 Organization size by number of users broken out by whether they have an incident response plan addressing ransomware

	1–9		10–49		50–99		100–249		250–499		500–999		1000–4999		5000–9999		10,000+		Total
Yes	3	6%	3	6%	3	4%	2	4%	2	6%	3	0%	0	22%	11	8%	4	44%	22	50%
No	2	25%	2	12.5%	1	0%	1	0%	0	12.5%	1	0%	0	25%	2	0%	0	25%	2	8%

Table 3.7 Survey question asking whether respondents orgaization has a contract with a third party negotiator

Q8	Does your organization have a contract with a third party negotiator in case of the need to negotiate ransoms?		
#	Answer	%	Count
1	Yes	25.00%	9
2	No	75.00%	27
	Total	100%	36

Table 3.8 Survey question asking whether a respondents organization would pay a ransom in an extortion-based cyberattack

Q11	Would our organization pay a ransom for any of the following reasons?		
#	Answer	%	Count
1	To regain access to files that the company has no other way of accessing	20.97%	13
2	To restore operations even if good backups are available, because of the time required to restore operations	4.84%	3
3	The cost of the ransom is lower than the cost to restore operations	8.06%	5
4	To avoid publicity, preserve reputation, avoid bad press, etc.	12.90%	8
5	If there was no other choice because backups do not exist or work.	24.19%	15
6	No, I would never pay a ransom regardless of time, cost, or feasibility to recover.	25.81%	16
8	To receive a Certificate of Destruction stating that exfiltrated data has been destroyed.	3.23%	2
	Total	100%	62

Table 3.9 Survey question about scenarios in which an organization would pay a ransom

Q12	Would your organization pay a "reasonable" ransom demand if it would take longer than a certain amount of time to recover operations?		
#	Answer	%	Count
1	To recover operations in less than 24 h	10.34%	3
2	To recover operations in less than 48 h	0.00%	0
3	To recover operations in less than 72 h	10.34%	3
4	To recover operations in less than 7 days	3.45%	1
5	To recover operations in less than 14 days	0.00%	0
6	To recover operations in less than 28 days	0.00%	0
7	To recover operations in the case that the restore process fails	20.69%	6
8	No	55.17%	16
	Total	100%	29

and municipalities, among many others. Ransomware is more than a simple internet crime as its effects are potentially fatal to both businesses and people. The following sections cover the impact of ransomware on various sectors.

Small to Medium Sized Businesses (SMB)

SMBs are businesses with under 1000 employees. Companies with more than 1000 employees are considered "large businesses."[4] SMBs are particularly susceptible to cyberattacks because of their limited cybersecurity resources. Large organizations can afford to enlist teams of security people that run security operations and training. This is often not the case with SMBs as they often rely on Managed Service Providers (MSPs) or their IT person to run security. Businesses on the smaller end of the spectrum struggle to expand their security program beyond anti-virus and file permissions. One way for businesses to overcome this problem is by hiring "Managed Service Providers" (MSPs) to manage IT and security. MSPs provide dedicated IT and security services that replicate large IT departments to numerous customers in diverse industries. A 2019 survey of MSPs by Datto Inc. revealed the following[5]:

- 85% of MSP's reported ransomware attacks against SMB customers in the previous two years, with 56% occurring in 2019.
- 28% of SMBs are "very concerned" about ransomware
- 45% of MSP's reported "business threatening downtime" at their clients.
- The average cost of downtime was $141,000, while the average ransom request was $5900
- 92% of SMBs with Business Continuity and Disaster Recovery solutions were "less likely to experience significant downtime."
- 80% of SMBs with Business Continuity and Disaster Recovery solutions recovered in under 24 h.
- The median size for organizations targeted by ransomware in Q3 2020 was 168 users. (Coveware)

According to the 2020 Chainalysis Crypto Crime Report small-to-medium businesses are more likely to be targeted by Ransomware as a Service (RaaS) attacks than traditional ransomware.[6]

What can SMBs do?

The first thing that small businesses should do is define and inventory their most important assets.

- What is their most critical data?
- What data or systems could they not live without?
- Assess how the loss of availability to that data or systems would affect the company over the short, medium, and long term.

---

[4] Gartner (n.d.).
[5] Datto Inc. (2019).
[6] Chainalysis (2020).

This exercise will help quickly identify the organizations' most critical assets. There are many helpful guides on cybersecurity for small businesses from multiple US and international agencies.

Cyber guides from the US Government: https://www.fcc.gov/general/cybersecurity-small-business

Small Business Administration: https://www.sba.gov/business-guide/manage-your-business/stay-safe-cybersecurity-threats

FTC: https://www.ftc.gov/tips-advice/business-center/small-businesses/cybersecurity

NY State: https://www.nyssbdc.org/resources/cybersecurity.html

ENISA: https://www.enisa.europa.eu/publications/ransomware

Australia: https://www.cyber.gov.au/acsc/view-all-content/essential-eight

The risk of a business-ending ransomware incident is particularly high for SMBs because of their small budgets, limited staff, and easy connection to the internet. They need to defend themselves from external adversaries and "trusted" 3rd parties.

Managed Service Providers (MSPs)

Managed Service Providers supply IT services to organizations instead of or in complement with IT departments. They enable organizations to take advantage of powerful technologies but have become an increasingly dangerous risk because they have direct network access and the same technology flaws as any other business. The US National Cybersecurity and Communications Integration Center (NCCIC) issued a warning in October 2018 because of APT activity at MSPs.[7] APT stands for Advanced Persistent Threat. Typical APT activity is targeted, sophisticated, and long term. NCCIC warned that APT's were motivated by espionage and intellectual property theft. MSP's have become springboards for ransomware actors because they can compromise the MSP and all of its customers en masse.

ProPublica wrote an excellent article about MSP ransomware problems in late 2019. One of the stories they wrote was about a ransomware incident launched from an MSP covering dental businesses. The MSP went silent for a week after attackers launched the attack.[8] They said that they "are not optimistic about the chances for a full or timely recovery. At this time, we must recommend that you seek outside technical assistance with the recovery of your data."[9] The MSP announced 19 days later that they were shutting down. Copycat attacks are common in the ransomware industry. An attack on another MSP supporting dental practices resulted in 400

---

[7] US CISA (2020).

[8] Dudley, The New Target That Enables Ransomware Hackers to Paralyze Dozens of Towns and Businesses at Once (2019b).

[9] Ibid.

infected customers.[10] The MSP involved in that case obtained a recovery key. Another MSP compromise incident led to ransomware infection in 22 cities and towns, of which one town was victimized twice through the MSP.[11]

One of the problems with MSPs is that they are businesses like any other business and are subject to poor planning, cost-cutting, and unrealistic expectations, similar to any other organization. Huntress Labs is a US business with expertise in MSP ransomware. Kyle Hanslovan and Chris Bisnett from Huntress stated that MSPs become overwhelmed and pay the ransom in many cases.[12] Emsisoft is a UK-based company that helps organizations with ransomware incidents by looking for weaknesses in the code.[13] Fabian Wosar is their CTO. And online technology magazine CRN interviewed him where he said:

> Hacking an MSP and then encrypting all their clients is hugely profitable. There is such a huge return on investment … its low hanging fruit. MSPs never had to deal with it, so in a way they got away with a lot of shady practices, and bad cyber-hygiene. Either they were lazy or they didn't know any better, you had a lot of them who are vulnerable to this type of attack.[14]

Ransomware actors find compromising MSPs efficient because they negotiate with the MSP rather than their clients. It is often unclear who is responsible for paying the ransom, even if the victim has a cyber-insurance policy. There are documented cases of MSPs paying, asking their customers to pay, or going out of business.[15] The Texas Department of Information Resources released recommendations for controlling RDP, Tor traffic, Pastebin access, monitoring PowerShell, and controlling remote access software that limits the risks posed by MSPs. It is good advice that will help prevent ransomware outbreaks.[16]

There are things that companies can do when hiring MSPs:

- Look for MSPs with customers in a diverse set of industries.
- Put MSP's through a 3rd Party Risk Assessment process.
- Demand multifactor authentication for email, administrative access, and computer systems.[17]
- Determine in advance who is responsible for negotiating and paying ransoms, and include it in the contract.
- Negotiate the requirement for the MSP to carry cyber insurance.
- Don't give the MSP direct network access. Force them to use customer issued equipment and MFA authentication, similar to any other remote employee.

---

[10] Ibid.

[11] Ibid.

[12] Hanslovan and Bisnett, hack_it 2020 (2020).

[13] Johnson (2019).

[14] Ibid.

[15] Dudley, The New Target That Enables Ransomware Hackers to Paralyze Dozens of Towns and Businesses at Once (2019b).

[16] Texas Department of Information Resources (2019).

[17] Hicock (2016).

Education

Education at all levels is a target for ransomware because of the combination of heavy MSP use and less sophisticated technology practices that closely resemble local governments. 2019 was the "summer of ransomware" attacks on schools, with more than 500 schools being victimized.[18] The attacks continued into 2020. There was a 338% increase in ransomware events at schools in Q3 2020 over the previous quarter.[19] In 2019 there was a similar jump in the number of incidents, with a 1020% increase during the same period.[20] The severe increase in ransomware incidents made 2019 the summer of ransomware. Q3 includes July, August, and September. This is the exact time that schools are preparing to open, leading to delays in opening, lost data, and ransoms paid using both public and insurance funds.[21] Educational institutions always need to be on alert, but the beginning of the school year is an incredibly delicate time because there is new staff, new students, and unfamiliar technology. Ransomware operators like to exert maximum pressure, and there is no better time than at the beginning of the school year.

Thirty-one percent of Q3 2019 ransomware attacks against schools also included data exfiltration.[22] Ransomware is a problem in every industry, but schools hold sensitive personal and medical data about children long after leaving school. This was the case in September 2020 in the Washington D.C. suburb of Fairfax, Virginia, when the Maze group posted student and employee personal information on the internet as part of a ransomware attack.[23] This is not an issue just for parents. The children may grow up to find their "permanent records" posted on the internet.

Conversely, ransomware actors also threaten to delete school records to increase pressure. The Athens, Texas Independent School District faced this choice. They found the "permanent loss of many years' worth of records and delaying the start of school for many weeks"[24] to be a worse choice than paying the ransom. Some might find the loss of their "permanent record" appealing. Ultimately, it is up to parents and other community members to demand security improvements at schools and be willing to pay for them.

---

[18] Cimpanu, Over 500 US schools were hit by ransomware in 2019 (2019b).

[19] Emsisoft Malware Lab (2020).

[20] Ibid.

[21] Wellerman and Morning (2020).

[22] Emsisoft Malware Lab (2020).

[23] Security Magazine (2020).

[24] Wellerman and Morning (2020).

Healthcare

The healthcare sector and hospitals, in particular, are frequent victims of ransomware. They have large computer networks with facilities partially open to the public, a wide variety of systems, large numbers of medical IoT devices, extremely sensitive data, and often urgent work. Ransomware actors like to hide in the noise, strike at the worst time (for the victim), and exert maximum pressure. There is hardly a worse mix of factors than that of the healthcare industry. Consequentially, attacks have been numerous, especially since the beginning of the COVID-19 Coronavirus pandemic in March 2020. They have been cited as the cause for increased heart attack deaths because of delayed care, averaging patients 2.7 min longer to receive an electrocardiogram.[25] The first death directly attributed to ransomware occurred in Germany in September 2020 because an attack forced the hospital to refuse new patients.[26] Attacks are not limited to human healthcare. Ryuk ransomware hit four hundred veterinary hospitals for the 2nd time in November 2019.[27]

The first widely reported worm-like ransomware infection in the healthcare industry was the 2017 WannaCry outbreak. It had significant impacts on the UK National Health Service with widespread computer infection and disruption of organizations, trusts, and primary care facilities.[28] This disruption had a damaging impact on healthcare delivery by forcing the cancellation of an estimated 19,000 appointments, including surgeries.[29] The outbreak forced the routing of emergency care patients to other hospitals. There are no reports of increased deaths in either the original UK National Audit Office report or subsequent Nature Partner Journal research.[30] Nevertheless, WannaCry made the concept of wormable disruptive ransomware a boardroom level discussion.[31] Unfortunately, the use of ransomware has worsened and led to deadlier outcomes since 2017.

The year 2020 saw the collision of brutally efficient ransomware gangs with a pandemic leading to government warnings about cyberattacks on healthcare entities in both the US and the UK. That was for a good reason. According to the 2020 Verizon DBIR, the number of reported breaches increased to 521 in the 2020 report from 304 in the 2019 report.[32][33] The breaches included both personally identifiable information (PI) and protected health information (PHI).

Multiple attacks against healthcare providers in mid to late 2020 caused the US CISA, FBI, and DHS to release an alert about the "imminent threat" of ransomware

---

[25] Akpan (2019).

[26] Goodin, A Patient Dies After a Ransomware Attack Hits a Hospital (2020).

[27] Krebs, Ransomware Bites 400 Veterinary Hospitals (2019).

[28] UK Natoinal Audit Office Comptroller and Auditor General (2018).

[29] Ibid.

[30] Ghafur et al. (2019).

[31] Seals (2018).

[32] Langlois (2020).

[33] Widup (2019).

activity.[34] This alert followed a Ryuk ransomware attack against all US facilities of a US based company called Universal Health Services (UHS). The attack struck over 400 medical facilities and is notable for its many impacted facilities and the employees' response.[35] Some of the UHS's 900,000 employees took to Reddit and other platforms to report a likely ransomware incident, which indicates that users require training on handling incidents.[36] Another interesting aspect of this attack is that a cybersecurity intelligence company called Advanced Intel claims to have detected Emotet and Trickbot use at UHS "throughout 2020," which means that threat actors were on the network for an extended time before launching the final attack.[37]

There are resources available to assist healthcare providers in hardening their prevention, detection, and response capabilities. (Appendix D)

- 2020 CISA/FBI/HHS Joint Cybersecurity Advisory: Ransomware Activity Targeting the Healthcare and Public Sector
- 2020 CISA, MS-ISAC Ransomware Guide
- HHS FACT SHEET Ransomware and HIPAA
- HHS HC3 Products. This is a great source of TTP information that is useful across sectors
- HICP: Health Industry Cybersecurity Practices
- H-ISAC: Health ISAC
- CIS: No-Cost Malicious Domain Blocking and Reporting for U.S. Hospitals

Laws and Regulations

The following section is written based on research and opinions at the time of writing. This report does not provide legal advice and should not be relied upon when evaluating the propriety of any action. Readers should consult a qualified attorney for up to date authoritative information.

Abstract
This section examines laws and financial regulations that directly and indirectly apply to ransomware. It also covers proposed legislation. Part of that coverage includes an interview with a NY State Senator who proposed outlawing the use of public funds to pay ransoms.

Legal topics related to ransomware include the following:

- Cybercrime laws
- Extortion, conspiracy, and wire fraud

---

[34] US CISA, FBI and DHS (2020).

[35] Davis (2020).

[36] Ibid.

[37] Gatlan, UHS restores hospital systems after Ryuk ransomware attack (2020b).

- Ransom laws
- Terrorism financing laws
- Regulations, particularly regarding breach reporting
- Privacy laws
- Ransomware bills, particularly regarding outlawing the use of public funds to pay ransoms

Civil litigation is not within this paper's scope but should be on the list of postbreach concerns for organizations. For example, a company called Blackbaud stopped a ransomware attack during the data exfiltration phase, but before encryption occured. They announced in their 3rd Quarter SEC filing (11/3/2020 10-Q) that they face 23 class-action lawsuits, and some of their customers have been sued directly as well.[38, 39]

Cybercrime Laws

The unauthorized use of ransomware is against national cybercrime laws globally. The Computer Fraud and Abuse Act (18 US Code § 1030) is the federal law covering cybercrime in general in the US.[40] In Europe, the 2013 Directive on "Attacks against Information Systems" provides the minimum recommended cybercrime penalties.[41] Neither legislation explicitly mentions "ransomware," although both are clear that unauthorized manipulation of computer systems is a crime. There are additional penalties for particular circumstances (critical infrastructure, government systems, or the involvement of organized crime).

The US government is willing to levy a raft of charges against ransomware operators and other cybercriminals if they can identify them. For example, a North Korean programmer named Park Jin Hyok was charged with numerous crimes for his involvement in the WannaCry ransomware outbreak, as well as other attacks.[42] However, many of the people involved in ransomware operations are outside the US or are in places that ignore cybercrime activity as long as the victims are outside the host country.[43] The US Department of Justice claims that it will "pursue cybercriminals for as long as it takes," and that was the case for Nathan Wyatt of the United Kingdom.[44] He took a manual approach to the ransomware game. He obtained

---

[38] Gatlan, UHS restores hospital systems after Ryuk ransomware attack (2020b).

[39] Alder (2021).

[40] 18 U.S. Code § 1030—Fraud and related activity in connection with computers (n.d.).

[41] European Parliament (2013).

[42] US Department of Justice (2018).

[43] Palmer, Ransomware: Big paydays and little chance of getting caught means boom time for crooks (2019a).

[44] US District Court, Eastern District of MO St. Louis (2017).

stolen personal information from multiple businesses.[45] He then contacted people at the businesses with evidence that he possessed stolen information and a threat to release it to the public if they did not pay $75,000–$350,000 in Bitcoin. Wyatt was charged with multiple identity theft, conspiracy, and computer crimes. He was extradited to the US, where he pleaded guilty, was sentenced to five years in federal prison, and was ordered to pay $1,467,048 in restitution.[46] The two cases demonstrate that law enforcement will attempt to identify cybercriminals and that charges against them extend to non-cybercrimes such as extortion, wire fraud, and conspiracy.

Paying Ransoms Is Legal. Paying Ransoms to Terrorists Is Not

One of the most surprising aspects of the ransomware situation is that paying a ransom is generally legal, whether for people, property, or data.[47, 48] (It is illegal for family members to pay ransoms for kidnap victims without prosecutor permission in Italy.)[49] There are moral, ethical, and precedential issues with paying ransoms, but there are very few legal issues for those paying. This is true both in the US and Europe.[50] The problem comes into play when the payment could be directed to sanctioned people or groups and terrorism.

The US Department of Treasury, Office of Foreign Assets Control (OFAC) made headlines in October 2020 when it announced that paying ransoms to sanctioned entities for ransomware poses the risk of "civil penalties."[51] OFAC describes itself as follows:

> The Office of Foreign Assets Control administers and enforces economic sanctions programs primarily against countries and groups of individuals, such as terrorists and narcotics traffickers. The sanctions can be either comprehensive or selective, using the blocking of assets and trade restrictions to accomplish foreign policy and national security goals.

On the cyber front, OFAC is responsible for maintaining the list of sanctioned entities and granting licenses to people or organizations that want to send them money. The key facts about the sanctions process are:

- OFAC, along with the US Department of the Treasury and US Department of State, designate entities and people for sanctions.[52]

---

[45] Ibid.

[46] US Department of Justice (2020a, b, c).

[47] Dutton and Bellish (2014).

[48] Huffman et al. (n.d.).

[49] Bohlen (1998).

[50] Worlock (2020).

[51] US Department of The Treasury, Office of Foreign Assets Control (OFAC) (2020).

[52] US Department of The Treasury (2002).

- Sanctioned entities are placed on a "Specially Designated Entities" (SDN) list. The lists are publicly available, searchable, and communicated via press releases.[53], [54], [55] An example of a "cyber-related" sanction announcement from September 23, 2020, can be found here: https://home.treasury.gov/policy-issues/financial-sanctions/recent-actions/20200923_33

Once an entity is added to a sanction list, organizations "risk violating OFAC regulations" if they pay a ransom to that entity. The fact that organizations "risk violating" sanctions regulations is important because OFAC publicly announces enforcement actions. A ransomware victim might pay a ransom to avoid negative publicity related to an incident, only to receive worse publicity when the government announces punitive action over a regulatory violation. It is not a ban on paying ransoms, nor does it make paying ransoms illegal, as some of the press has stated.

I contacted OFAC on 12/22/2020 to clarify what is and is not permitted. OFAC referred to the 10/1/2020 Ransomware Advisory and answered several questions.[56]

Question to OFAC Should an OFAC licensing application be filed when it is unknown that the payment is to a sanctioned entity? What if the person or entity is known to not be on the sanctioned list?

OFAC Response *Companies can apply for a license if they seek to engage in a transaction/activity that is otherwise prohibited.*[57]

However, as stated in the advisory, license applications will be reviewed by OFAC on a case-by-case basis with a presumption of denial. If the company knows that the transaction/activity does not involve a sanctioned party/jurisdiction, there may not be a reason to contact OFAC unless you have questions or a sanctions nexus is later discovered. However, we recommend that you document how you came to the conclusion that there isn't a sanctions nexus and we still recommend that you work with law enforcement before responding to the ransom attack. Law enforcement will be able to help you determine whether or not you need to contact OFAC.

In instances where a sanctions nexus is known, please contact OFAC via this email address, OFAC_Feedback@treasury.gov, and provide us with the following information:

1. *Company or individual name, as well as any other identifying information, for the subject(s) targeted by a ransomware attack or involved in facilitating the ransomware payment; clarify whether any of the foregoing are U.S. persons or otherwise subject to U.S. jurisdiction.*

---

[53] US Department of The Treasury (n.d.-a, -b, -c).

[54] OFAC (n.d.).

[55] US Department of The Treasury (n.d.-a, -b, -c).

[56] US Department of The Treasury, Office of Foreign Assets Control (OFAC) (2020).

[57] OFAC_Feedback@treasury.gov (2020).

2. *Type of ransomware variant, if available.*
3. *Key dates and facts, including the date of attack, date the attack was discovered, ransomware payment deadline, payment instructions, and ransomware payment amount demanded.*
4. *An explanation and evidence, if available, of any sanctions nexus.*
5. *A statement whether the ransomware attack has been reported to law enforcement, and, if so, which law enforcement entities are aware of the incident.*

Question to OFAC Are OFAC licensing applications public information? Can anyone request to see them? Are they stored openly on the internet or is there a request process?

OFAC Response *Specific licenses are not publicly available.*[58]

OFAC's answer did not directly address this question. High-profile license applications have been made publicly available through FOIA in the past. OFAC even keeps a list of "frequently requested" applications on its website, including several for Beyonce and Jay-Z's 2013 Cuba visit.[59] Organizations should be aware that it is theoretically possible for OFAC license applications to be retrievable via the Freedom of Information Act (FOIA) process.

Question to OFAC Have there been any enforcement actions for sanctions violations related to Ransomware payments?

OFAC Response *You can view all of OFAC's enforcement actions here: https:// home.treasury.gov/policy-issues/financial-sanctions/recent-actions/1321*[60]

Organizations considering paying ransoms for themselves or others must seek legal advice while closely following OFAC regulations and other US laws. US sanctions potentially have a global impact because of the extraterritorial reach of US laws. Compliance might also be required of organizations outside of the US, depending on where they conduct business. There is a risk that ransomware victims inadvertently pay organizations on US terrorism lists.[61] Terror funding restrictions are not limited to the US. The UK, Wales, and others have similar laws.[62]

The consequence of sanctions violations is civil penalties. The October 2020 OFAC Ransomware advisory states that there are two things organizations can do to mitigate potential penalties.[63]

---

[58] Ibid.

[59] US Department of The Treasury (n.d.-a, -b, -c).

[60] OFAC_Feedback@treasury.gov (2020).

[61] Wilkie Compliance (n.d.).

[62] Simmons + Simmons (2018).

[63] US Department of The Treasury, Office of Foreign Assets Control (OFAC) (2020)

- "The existence, nature, and adequacy of a sanctions compliance program is a factor that OFAC may consider when determining an appropriate enforcement response (including the amount of civil monetary penalty, if any)."
- "Self-initiated, timely, and complete report of a ransomware attack to law enforcement to be a significant mitigating factor in determining an appropriate enforcement outcome if the situation is later determined to have a sanctions nexus."

There is no evidence that an OFAC enforcement action has been taken against ransomware victims as of the time of this writing.

There are additional financial regulations by the US Financial Crimes Enforcement Network (FinCEN) for those involved in "facilitating payments." FinCEN, like OFAC, is part of the Department of Treasury. Their duties include identifying money laundering and creating financial intelligence reports. They also have an investigative arm that collaborates with the Department of Justice.[64] Activities such as money laundering using cryptocoin mixers and other techniques is within the purview of FinCEN. FinCEN developed indicators of ransomware activity that it monitors to identify suspicious transactions.[65]

Financial institutions, ransomware, and extortion victims should be aware that "entities engaged in money services business activities" are required to file Suspicious Activity Reports (SARs) for related transactions of $5000 or more (with lower value exceptions in some cases).[66]

The US Securities and Exchange Commission (SEC) recommends that organizations disclose "material" cybersecurity risks and incidents via the 8-K, 6-K, 10-K, and 20-F forms.[67] The SEC's recommendations center on transparency to investors. They refer to their guidance as recommendations but also say, "It is critical that public companies take all required actions to inform investors about material cybersecurity risks and incidents in a timely fashion."[68]

The interweaving nature of US laws, regulations, and enforcement activities demonstrates that organizations must include their legal, compliance, and finance departments in Incident Response plans.[69] Additionally, organizations must practice cyberattack response with all groups to learn how to work through complex requirements on short notice while under pressure.

---

[64] US Department of Justice (2020a, b, c).

[65] FinCEN (2020).

[66] Ibid.

[67] US Securities and Exchange Commission (2018).

[68] Ibid.

[69] Ramachandran et al. (2020).

Privacy Laws

The topic of privacy laws boils down to whether a privacy law considers any aspect of a ransomware incident a reportable data breach. Major privacy laws typically outline what constitutes protected personal information, a security breach, and a reportable data breach. They cover other areas such as using and selling personal information, children's privacy, and appropriate safeguards for data. These topics all feed into the ransomware topic, but are not directly within the scope of this paper. However, organizations should remember that attackers cannot ransom personal information that organizations do not collect and retain. This section covers what organizations need to know about privacy laws related to ransomware.

Organizations should seek guidance from privacy professionals to determine which laws apply to them. National or regional laws can apply depending on where the data subject resides (e.g. GDPR, LGDPA, CCPA, POPIA, and others). In other cases, laws apply because of the type of data being collected (HIPAA, COPPA, NY DFS, and state laws). These laws specify the acceptable collection and use of personal information. This is very important on a day-to-day basis, as any mismanagement of personal data will come to light in the event of a reportable data breach.

One of the most important aspects of ransomware response is caution with the word "breach" because it is a legally defined term in most privacy and cybersecurity laws.[70] The declaration of a "data breach" may require public disclosure. An example of this is with the GDPR.

Article 33.1 states:

> In the case of a personal data breach, the controller shall without undue delay and, where feasible, not later than 72 hours after having become aware of it, notify the personal data breach to the supervisory authority competent in accordance with Article 55, unless the personal data breach is unlikely to result in a risk to the rights and freedoms of natural persons.[71]

The Article 29 Data Protection Working Party report specifically examines when ransomware causes a reportable data breach. The report provides several examples that the Working Party summarizes in this table (Fig. 3.1):

HIPAA is a US healthcare privacy law. The US Department of Health and Human Services released a document on ransomware in FAQ format called FACT SHEET: Ransomware and HIPAA. The document states that a "breach" is "presumed to have occurred" unless the victim can demonstrate a "low probability that Personal Health Information (PHI) has been compromised."[72] It goes on to explain the steps required to prove the low probability of compromise.

Two key takeaways for organizations are that they must understand privacy laws before an incident and that using the word "breach" has legal effects that require

---

[70] Freedman, Security Economy Episode 3: Got a Privacy Plan? GDPR, CCPA, and the Rise of Ransomware (2020b).

[71] European Parliament (n.d.).

[72] US Department of Health and Human Services (n.d.).

assessment. An incident is not necessarily a breach under applicable laws, and an assessment of the facts of each incident is critical to determine whether notification about the incident is required.[73]

Privacy laws offer attackers an additional avenue for extortion. A recent criminal tactic called "GDPR blackmail," threatens to inform regulators of a ransomware incident if organizations do not pay.[74] They even threaten to do this in their ransom note:

> All your data is a backed up. You must pay 0.015 BTC to 13JwJDaU3xdNFfcSySFCy95 E2Tko18fiyB 48 hours for recover it. After 48 hours expiration we will leaked and exposed all your data. **In case of refusal to pay, we will contact the General Data Protection Regulation, GDPR and notify them that you store user data in an open form and is not safe. Under the rules of the law, you face a heavy fine or arrest** and your base dump will be dropped from our server! You can buy bitcoin here, does not take much time to buy hxxps[://]localbitcoins[.]com with this guide hxxps[://]localbitcoins[.]com/guides/how-to-buy-bitcoins After paying write to me in the mail with your DB IP: restore_base@tuta[.]io.[75]

The Bitcoin Abuse Database shows that this ransom was likely paid: https://www.bitcoinabuse.com/reports/13JwJDaU3xdNFfcSySFCy95E2Tko18fiyB

Example	Notify the supervisory authority?	Notify the data subject?	Notes/recommendations
iv. A controller suffers a ransomware attack which results in all data being encrypted. No back-ups are available and the data cannot be restored. On investigation, it becomes clear that the ransomware's only functionality was to encrypt the data, and that there was no other malware present in the system.	Yes, report to the competent supervisory authority, if there are potential consequences to individuals as this is a loss of availability.	Yes, report to individuals, depending on the nature of the personal data affected and the possible effect of the lack of availability of the data, as well as other likely consequences.	If there was a backup available and data could be restored in good time, this would not need to be reported to the supervisory authority or to individuals as there would have been no permanent loss of availability or confidentiality. However, the supervisory authority may consider an investigation to assess compliance with the broader security requirements of Article 32.

Fig. 3.1 GDPR Article 29 Working Party "Examples of personal data breaches and who to notify." (Source: (ARTICLE 29 DATA PROTECTION WORKING PARTY 2017))

---

[73] Freedman, Security Economy Episode 3: Got a Privacy Plan? GDPR, CCPA, and the Rise of Ransomware (2020b).

[74] Abrams, Surge of MongoDB ransom attacks use GDPR as extortion leverage (2020e).

[75] Ibid.

GDPR is not the only privacy law used in these threats, but it is the most well-known. Attackers seek to apply maximum pressure. Organizations might feel that they can recover all their data, but they have to consider whether they can also tolerate a publicly disclosed data breach. Attackers know that breach response is likely more expensive than a ransom demand. Using these kinds of techniques is known as "double extortion," which is the combination of two extortion techniques.[76] (Ransomware + threaten to leak data, ransomware + threaten to inform regulators).

(Attempted) Ransomware Laws

There are currently two types of ransomware laws. Most fall into the "computer crimes" category of unauthorized use of ransomware. Computer crime laws sometimes explicitly refer to ransomware while implying it in other cases. The US Computer Fraud and Abuse Act (18 US Code § 1030) covers ransomware implicitly in two different ways. The first is unauthorized damage to a protected computer[77]:

18 US Code § 1030 (a)(5)-
(A) knowingly causes the transmission of a program, information, code, or command, and as a result of such conduct, intentionally causes damage without authorization, to a protected computer;
(B) intentionally accesses a protected computer without authorization, and as a result of such conduct, recklessly causes damage; or
(C) intentionally accesses a protected computer without authorization, and as a result of such conduct, causes damage and loss.

18 US Code § 1030 (5) is a US federal law that is similar to other computer crime laws at the federal and state levels. It also has an extortion clause that appears in some state laws:[78]

18 US Code § 1030 (a)(7) Whoever-
(7) with intent to extort from any person any money or other thing of value, transmits in interstate or foreign commerce any communication containing any--

> (A) threat to cause damage to a protected computer;
> (B) threat to obtain information from a protected computer without authorization or in excess of authorization or to impair the confidentiality of information obtained from a protected computer without authorization or by exceeding authorized access; or

---

[76] Check Point Research (2020).
[77] Congressional Research Service (2020).
[78] Ibid.

(C) demand or request for money or other thing of value in relation to damage to a protected computer, where such damage was caused to facilitate the extortion.

Most passed and proposed ransomware laws are stylistically similar to the CFAA. (Appendix E) A few states have attempted to outlaw ransom payments using public money. The mention of public money is important. It does not apply to companies or individuals, and it leaves a loophole for insurance policies purchased by public entities. Two of these bills have passed as of the time of this writing. The first was in North Carolina. N.C. General Statute §143–800(a) states:

> No State agency or local government entity shall submit payment or otherwise communicate with an entity that has engaged in a cybersecurity incident on an information technology system by encrypting data and then subsequently offering to decrypt that data in exchange for a ransom payment.[79]

Notably, that the law includes communicating with threat actors, making it impossible for government entities to communicate directly.

Florida is the only other state that has passed a law preventing payments as of March 2024. Florida's law, passed July 1, 2022 is narrower in scope since communication is not prohibited. Florida Statute §202.3186 states:

> A state agency as defined in s. 282.318(2), a county, or a municipality experiencing a ransomware incident may not pay or otherwise comply with a ransom demand.[80]

There are other proposed bills. The National Conference of State Legislatures provided the information contained in the following section.

Georgia H.B. 1133: "Relates to general provisions of state government so as to prohibit state agencies from paying ransoms in response to cyber-attacks."[81]

Status: Failed-Adjourned

Iowa S.B. 2391 (renumbered from Iowa S.B. 2080): "Prohibits the state and political subdivisions of the state from expending public moneys for payment to persons responsible for ransomware attacks."[82]

Status: Failed-Adjourned

New York S.B. 7289: Prohibits any municipal corporation or other government entity from paying ransom in the event of a cyber-attack against such municipal corporation's or government entity's critical infrastructure.[83]

Status: Pending

---

[79] North Carolina General Assembly (2021).

[80] Florida State Legislature (2022).

[81] National Conference of State Legislatures (2020).

[82] Ibid.

[83] Ibid.

New York S.B. 7246: Creates a cyber security enhancement fund to be used for the purpose of upgrading cyber security in local governments, including but not limited to, villages, towns and cities with a population of one million or less **and restricts the use of taxpayer moneys in paying ransoms in response to ransomware attacks.** (Full Text in Appendix F).

Status: Pending

New York S.B. 7246 was sponsored by State Senator Phil Boyle.[84] I contacted Senator Boyle, and his office agreed to an interview with the senator on October 1, 2020. The full edited transcript is in Appendix G.

When I Senator Boyle asked why he proposed the legislation, he said, *"As a fiscal conservative, it was driving me crazy that taxpayer money was going towards paying these ransoms."*[85] What makes his proposal unique is that it gives municipalities time and funding to make upgrades before outlawing the use of public funds. I asked what he expected from his bill if it became law. He said that it would "open a discussion" and raise awareness with citizens who are unaware of the problem.

We then discussed whether he felt it was acceptable to use insurance funds from policies purchased with public funds. He said he did not have a problem with that, even if there was a premium cost. This could be seen as a loophole while also reflecting the need for municipalities to transfer risk.

The next question was motivated by the fact that ransomware pivoted toward extortion due to the threat of data leakage. Ransomware initially started as a "locker" type software that encrypted files without exfiltrating them. It evolved into stealing data before encrypting. This put additional pressure on organizations to pay ransoms to have the data "destroyed." Ransomware operators know that obtaining a certificate of destruction can prevent a ransomware incident from becoming a reportable data breach. The release of private data is damaging to any organization, especially governments and school districts. For example, hackers stole student records from the Fairfax, VA school district and posted them on the internet. I asked Senator Boyle if it was preferable to pay ransoms rather than having data posted on the internet. He said that the public entities' leader would have to make that call, but he does not trust criminals to keep their word and destroy their copies. Deciding whether to trust criminals to keep their word and act in good faith is one of the most challenging decisions in the ransomware response process. It is a risk-based decision that all organizations should consider in advance during the incident preparation process.

We also discussed whether data loss in cases where data is not recoverable could lead to a loss of government trust. Senator Boyle emphatically stated that it would lead to a loss in trust and gave the example of the 1973 Veterans Affairs National Personnel Records Center (NPRC) fire.[86] The fire destroyed 80% of Army records

---

[84] The New York State Senate (n.d.).

[85] Boyle, NY State Senator (2020a).

[86] US Department of Veteran Affairs (n.d.).

from 1912 to 1959 and 75% of Air Force records from 1947 to 1963. Senator Boyle gave the example of World War II veterans who could not prove that they served in the military. The American Archivist magazine performed a case study of the NPRC fire and the insights are illustrative of the challenges of data loss. The case study is called "The National Personnel Records Center Fire: A Study in Disaster" and is found on the US National Archives and the American Archivist website (Volume 37, Issue 4).[87, 88]

Senator Boyle astutely described the challenges that societies face with ransomware. It is ethically questionable to fund criminal activity, while trusting criminals to keep their word. This puts governments in a difficult position with their constituents. IBM and Morning Consult conducted a survey on ransomware and reported that 56% of Americans disapprove of local governments using tax money to pay ransoms.[89] Conversely, 54% of Americans opposed paying additional taxes to protect their municipality from attacks.[90] When asked about the statistics, Senator Boyle said "that paying a little bit now for prevention is worth paying a lot down the road when the cyber ransom attacks takes place," thus justifying his proposal for a cybersecurity fund.

Overall, governments are in a difficult position with respect to ransomware. They need to balance the need for flexibility in responding to incidents with the potential outrage stemming from paying off criminals with tax money. This is likely why few bills outlawing ransom payments have passed up to early 2024.

---

[87] Stender and Walker, The National Personnel Records Center Fire: A Study in Disaster (1974a).

[88] Stender and Walker, The National Personnel Records Center Fire: A Study in Disaster (1974b) Reprint.

[89] Morning Consult + IBM Security (2019).

[90] Ibid.

Chapter 4
Preparation, Response, and Recovery

Contents

Abstract This chapter fully explores the challenges and strategies surrounding ransomware incidents. This chapter examines the crucial roles of preparation, response, and recovery alongside the contentious role of cyber insurance. The insurance industry and conventional wisdom depict cyber insurance as a business safeguard, providing necessary financial coverage for data breaches and recovery costs. It also highlights its potential to inadvertently fuel ransomware attacks by making payments more feasible. This chapter underscores the complex decision-making process businesses face when confronted with ransom demands, where downtime costs often far exceed the ransom itself, as illustrated by significant incidents in Baltimore and Atlanta. The reader engages with the debate on whether to pay ransoms, influenced by pressures from management, insurance companies, and legal considerations, against a backdrop of governmental advisories discouraging such payments. A notable discussion on the tactical aspects of incident response, particularly "The Great Shut Down, Reboot, Pull the Plug debate," reflects the nuanced decisions that IT professionals must make to mitigate damage without compromising future recovery efforts. This analysis extends beyond the financial implications, touching on the technological and operational challenges of effectively responding to and recovering from ransomware attacks, including the importance of secure backups, the potential for repeat attacks, and the strategic improvement of security postures through insurance requirements. This multifaceted examination reveals the complex interplay between cybersecurity practices, insurance policies, and the evolving tactics of cybercriminals, highlighting the need for a balanced and informed approach to ransomware defense and recovery.

Cyber Insurance

Cyber insurance has become an accelerant in the ransomware game because it is the cyber equivalent to oxygen to a fire. It is necessary for survival and can be explosive when it is overused. Cyber insurance is necessary to cover data breach expenses, cyber extortion, investigation costs, lost revenue, recovery, and other expenses. Cyber insurance can be the difference between a relatively quick recovery and going out of business. Any business that possesses third-party data or otherwise cannot afford the potential downtime that ransomware and other cyber events bring should consider it. This is particularly true for medium-sized and large businesses.

The cyber insurance-ransomware problem is about the financial impact of downtime. Decryption is less time consuming than restoration. There is enormous pressure to restore operations during an incident. Some of the pressure is naturally from management, but some of it is also from insurance companies that know it is less costly for them to pay the ransom.[1, 2] Insurance companies often provide investigative and recovery support because they are interested in reducing payouts related to downtime.[3]

There is data comparing ransom costs vs. downtime costs. The ransom for the May 2019 Baltimore ransomware incident was $76,000.[4] Baltimore's budget office estimated that the incident would "cost at least $18.2 million," including three weeks of downtime.[5] Attackers hit Atlanta with ransomware in March 2018. The ransom demand was $51,000. Atlanta estimated that it cost up to $17 million to recover.[6] The FBI, CISA, and other government organizations say that victims should not pay ransoms.[7] The report also recommends that the best way to eliminate ransomware is to outlaw ransom payments. However, the reality is that the decision to pay is a business decision where the choice is between going out of business or paying off criminals. Victims are under pressure from all sides. One side is pressure to get back to work, while another is that insurance companies try to keep costs

---

[1] Dudley, The New Target That Enables Ransomware Hackers to Paralyze Dozens of Towns and Businesses at Once (2019b).

[2] Palmer, Ransomware: Cyber-insurance payouts are adding to the problem, warn security experts (2019b).

[3] Munshaw and Marshall (2019).

[4] Dudley, The New Target That Enables Ransomware Hackers to Paralyze Dozens of Towns and Businesses at Once (2019b).

[5] Raver (2019).

[6] Freed (2019).

[7] FBI (n.d.).

down. A third comes from the government trying to stop the ransomware epidemic by cutting off payments. In the end, survival overtakes benevolence for most corporate victims.[8]

The existence of cyber insurance can also make an organization a target. A typical ransomware attack tactic is spending time on the network looking for insurance documents and incident response plans.[9] They use this information to determine the size of the ransom request and predict how organizations will respond. There have even been cases where the ransomware gang contacted the insurance company before contacting the victim, such as in the 2019 Lake City, Florida ransomware incident.[10] It is crucial to protect insurance information, but that is easier said than done. Insurance information is public for municipalities. It is not a secret that the Florida Municipal Insurance Trust is the carrier for many Florida municipalities.[11] Additionally, the US Securities and Exchange Commission (SEC) recommends reporting "the costs associated with maintaining cybersecurity protections, including, if applicable, insurance coverage relating to cybersecurity incidents or payments to service providers."[12] That level of required transparency puts organizations at risk. Financial details about insurance coverage open the door to targeted attacks. Another avenue of exposure is insurance companies themselves. Fabian Wosar from Emsisoft told a story about how companies listed on an insurance carrier's website subsequently fell victim to ransomware.[13]

Insurance coverage is of course not all bad, as insurance carriers improve their customers' security posture by improving their own due diligence processes. Pre-insurance assessments are becoming more intensive, with measurable results. An insurance company called Corvus scanned networks as part of the quotation process.[14] The scan looked for exposed RDP servers and unpatched VPN gateways.[15] The scans led to a 65% reduction in ransomware claims for new policies.[16] Insurance companies' self-interest to avoid paying claims motivates them to improve their clients' security posture. Insurance companies cannot take steps themselves to improve clients' security, but they can use their leverage to strongly motivate clients to make improvements. They can provide cybersecurity "experts" before and after breach to improve security.

There are many things for organizations to be wary of when purchasing insurance, and one of the biggest is "act of war" exceptions. Food manufacturer Mondelez

---

[8] LaCroix (2019).

[9] Palmer, Ransomware: Cyber-insurance payouts are adding to the problem, warn security experts (2019b).

[10] Gibson and Laporte (2019).

[11] Florida League of Cities (n.d.).

[12] US Securities and Exchange Commission (2018).

[13] Dudley, The Extortion Economy: How Insurance Companies Are Fueling a Rise in Ransomware Attacks (2019a).

[14] Abrams, Cyber insurer's security scans reduced ransomware claims by 65% (2020a).

[15] Ibid.

[16] Ibid.

International suffered a NotPetya ransomware attack in 2017 with over $100 million in damages.[17] The US Government attributes NotPetya to Russian Intelligence.[18] Mondelez's insurance company refused to pay the claim, citing "act of war" exceptions.[19] Acts of War and Terror designations in insurance policies are consequential and problematic for ransomware victims. The designations mean that insurance might not cover the claim or risk potential sanctions violations by paying when sanctioned entities are involved. There is some hope on the insurance front. Mondelez made the claim under an addendum to their property insurance policy for "the malicious introduction of a machine code or instruction."[20] Jason Crabtree from QOMPLX recommends purchasing "cyber-specific policies" that eliminate "war and terrorism" exclusions as the way to ensure organizations are covered.[21] Paul A. Ferillo from McDermott, Will, & Emery, LLP goes further, saying that the "Big 3" insurance companies (AIX, Chubb, AXA XL) "will not play games with your coverage."[22]

There are two outstanding articles available for further reading on the complexities of cyber insurance and ransomware.

ProPublica: *The Extortion Economy: How Insurance Companies Are Fueling a Rise in Ransomware Attacks*

https://www.propublica.org/article/the-extortion-economy-how-insurance-companies-are-fueling-a-rise-in-ransomware-attacks

The National Law Review: *A Ransomware Attack Could Devastate Your Company. Will Your Insurance Cover It?*

https://www.natlawreview.com/article/ransomware-attack-could-devastate-your-company-will-your-insurance-cover-it

Should You Pay?

The answer to the "should you pay" question is a complex web of legal, financial, security, privacy, and customer service considerations. Victims of ransomware need to treat it as a business decision. This is the conclusion of many experts, and they are

---

[17] Lemos (2020).

[18] US Department of Justice (2020a, b, c).

[19] Ibid.

[20] Satariano and Perloth (2019).

[21] Lemos (2020).

[22] Ferillo (May 18, 2020).

right.[23, 24, 25, 26, 27, 28]Feelings, ethics, and morals regarding paying ransoms do not matter when the company's existence and the livelihoods of hundreds or thousands of people are at stake. This section discusses the complexities of that decision, but that should not overshadow that it is a business decision that should be considered with other alternatives.[29]

If you're in IT and offer restores as a solution, can your restores get the company back online in a reasonable time frame? What are your Recovery Time Objective (RTO) and Recovery Point Objectives (RPO)? How close can you get to the stated RTO/RPO? Are you going to be able to restore on infected servers, or will you have to purchase all new hardware while the old ones are tied up in investigations? Have restorations on a similar scale been tested? Systems are called "backup systems" because they are primarily used to back up data. However, backup is not the point. The purpose of backup systems is to restore. They should be called restore systems.[30] Raymond Blum from Google made this point in his "How Google Backs Up the Internet" presentation. Anyone reading this paper should watch that presentation because it explains how Google backed up and **restored** Gmail. It grants perspective into the scale and complexity of mass restores.

NYC Tech Talk Series: How Google Backs up the Internet: https://www.youtube.com/watch?v=eNliOm9NtCM

Information Technology (IT) professionals should remember that traditional backup, restore, and DR methodologies do not work well in ransomware incidents when stakes are high, and patience is low. IT professionals will contribute pieces of information to the pay/don't pay decision, but this is not ultimately their choice.

If you're in InfoSec and you're repulsed by the idea of decrypting files on systems that will likely remain compromised and will definitely remain vulnerable, you have a point, but it's not ultimately your decision. The point of business is to be in business. InfoSec wants to improve security, so this never happens again, but that probably will not happen until the business is back online. Ransomware incidents are great opportunities to take steps toward betterment (read Recovery section on page 88).[31] This is unlikely to occur during an incident unless management is exceedingly patient or the incident is catastrophic but recoverable, and there is financial support to do so.[32]

---

[23] J. Munshaw (2019).

[24] Freedman, Ransomware: To Pay or Not to Pay (2020a).

[25] LaCroix (2019).

[26] Lovejoy (2020).

[27] Dudley, The Extortion Economy: How Insurance Companies Are Fueling a Rise in Ransomware Attacks (2019a).

[28] Dignan (2019).

[29] Zelonis (2019).

[30] Blum (October 22, 2013).

[31] Kudale (2020).

[32] Cimpanu, Ransomware incident to cost Danish company a whopping $95 million (2019c).

If you are a Privacy professional and do not want to pay the ransom because you cannot trust criminals to delete data, you have a point, but it is not your decision either. Privacy has a vital role in assessing whether a ransomware incident constitutes a reportable data breach, as well as the risks of potential further disclosure of data. Privacy may have a say in whether a ransom is paid because laws vary on what a reportable data breach is. Privacy has to determine whether paying the ransom to acquire a "certificate of destruction" certifying that attackers destroyed stolen data qualifies as a reportable data breach. Paying a ransom to "cover-up" what happened can lead to regulatory action, fines, and possibly criminal prosecution.[33] Privacy is one of the areas where morals, ethics, and laws matter the most. Companies should do the right thing for their customers and employees while following laws. Paying ransoms is a business decision as of 2021. Following the law is not. The incident response company Coveware recommends spending money on privacy attorneys rather than on ransoms.[34]

If you are a lawyer, it might be your decision. Legal is typically in the group of decision-makers that authorizes a ransom payment. Legal departments are in a challenging position with respect to ransomware. They have to ensure that they follow laws while protecting the company and customers under stressful conditions. A lot of the ransomware decision-making process is dumped on them because they evaluate the legality of payment, impacts on customers, employees, and suppliers, handle breach notifications, law enforcement notification, and possibly sanctions compliance and respond to subsequent enforcement actions. Ransomware victims should always notify outside counsel when considering paying a ransom.[35] Outside counsel provides professional expertise without emotion, pressure, and bias that is part of ordinary company politics.

If you are in finance, it might be your decision. The CFO/finance lead is in the group of people deciding whether to pay a ransom. Ransomware operators will take steps to exert maximum pressure. The "big game" ransomware operations likely know your ability to pay better than you do and will scale the ransom to make it affordable (in their estimation).[36] Finance will inform management of how the incident impacts the company's finances and forecast results based on different scenarios.

If you are the CEO, it is your decision. The CEO, owner, president, or designated leader makes the final decision on whether to pay a ransom. They rely on their leadership team to make the final decision in combination with law enforcement and the insurance carrier. It does not matter how unsavory or uncomfortable the decision is. As people of authority, they have to decide. That decision could involve paying a

---

[33] BBC News (2020).

[34] Krebs, Why Paying to Delete Stolen Data is Bonkers (2020c).

[35] Plesco and Shelhorse (2020).

[36] Dudley, The Extortion Economy: How Insurance Companies Are Fueling a Rise in Ransomware Attacks (2019a).

ransom or shutting down and laying off hundreds, as happened to The Heritage Company in December 2019.[37]

Governments will tell you not to pay. The FBI says:

> The FBI does not support paying a ransom in response to a ransomware attack. Paying a ransom doesn't guarantee you or your organization will get any data back. It also encourages perpetrators to target more victims and offers an incentive for others to get involved in this type of illegal activity.[38]

Europol also advises, "If you are a victim, do not pay the ransom!"[39] Europol backed nomoreransome.org says:

> Paying the ransom is never recommended, mainly because it does not guarantee a solution to the problem. There are also a number of issues that can go wrong accidentally. For example, there could be bugs in the malware that makes the encrypted data unrecoverable even with the right key.
>
> In addition, if the ransom is paid, it proves to the cybercriminals that ransomware is effective. As a result, cybercriminals will continue their activity and look for new ways to exploit systems that result in more infections and more money on their accounts.[40]

There are other considerations beyond the emotionless government statements. The following points illuminate the risks of paying ransoms.

- It might be better to let restores take their time and spend the money on security improvements.[41]
- There are documented cases of data leaking after paying the ransom.[42]
- Victims "re-extorted" over the same data within weeks of paying.[43]
- The decryption key does not work on some or all systems.[44]
- Average downtime due to ransomware in Q3 2020: 19 Days[45]
- 33% of "firms" do not get their data back. (2016 Trend Micro)[46]
- There are documented cases of the same firm being hit multiple times in a short period.[47]

---

[37] Cimpanu, Company shuts down because of ransomware, leaves 300 without jobs just before holidays (2020b).

[38] FBI (n.d.).

[39] EUROPOL (2020).

[40] No More Ransom (n.d.).

[41] Freedman, Ransomware: To Pay or Not to Pay (2020a).

[42] Krebs, Why Paying to Delete Stolen Data is Bonkers (2020c).

[43] Coveware (2020).

[44] Ibid.

[45] Ibid.

[46] Leyden (2016).

[47] Osborne (2020).

- It is difficult to trust recovered data previously encrypted by ransomware.[48] This is particularly true in a healthcare setting.[49] There is no evidence that ransomware was used as a cover to modify the contents of files, but it is theoretically possible.
- Even if victims attain a decryption key, it can contain bugs, takes time to run, and requires people to oversee the process.[50]

It is important to remember that ransomware operators are criminals and incidents are the result of deceitful acts. Ransomware gangs are just as untrustworthy as any other criminal gang. There are documented cases of hostage takers and pirates receiving ransom money and murdering hostages. Ransomware criminals cannot be trusted with company data, just as hostage-takers cannot with human lives.

The following graphic from Forrester Research is good for visualizing the scenario (Fig. 4.1).

Whether to pay or not is one of the most controversial and misunderstood topics in security. The controversy and misunderstanding feed into each other. That has led to a myth that victims do not need to pay if they have backups. The answer to the question of "should you pay?" is "it is complicated," but preparation makes the decision easier.

Response and Recovery

Providing comprehensive response plans is not within the scope of this paper. This type of advice exists from many sources, such as the high-quality CISA guide. Instead, the following section contains suggestions based on experience and examines some of the more debated ransomware incident response topics. Appendix H contains a diverse set of incident response resources from governments and the private sector.

The First 24 Hours

The first 24 h can be some of the most chaotic and crucial in a ransomware incident. This is often a time of uncertainty because alerts might be triggered, users might call about disruption to their work, ransom notes may appear, or other disruptive events may occur.

Don't Panic: That is probably the top advice in all incident response situations. Hopefully you have read this paper, created an incident response program, and practiced it many times. However, if not, rely on your training and experience. Document

---

[48] J. Munshaw (2019).

[49] Ibid.

[50] Zelonis (2019).

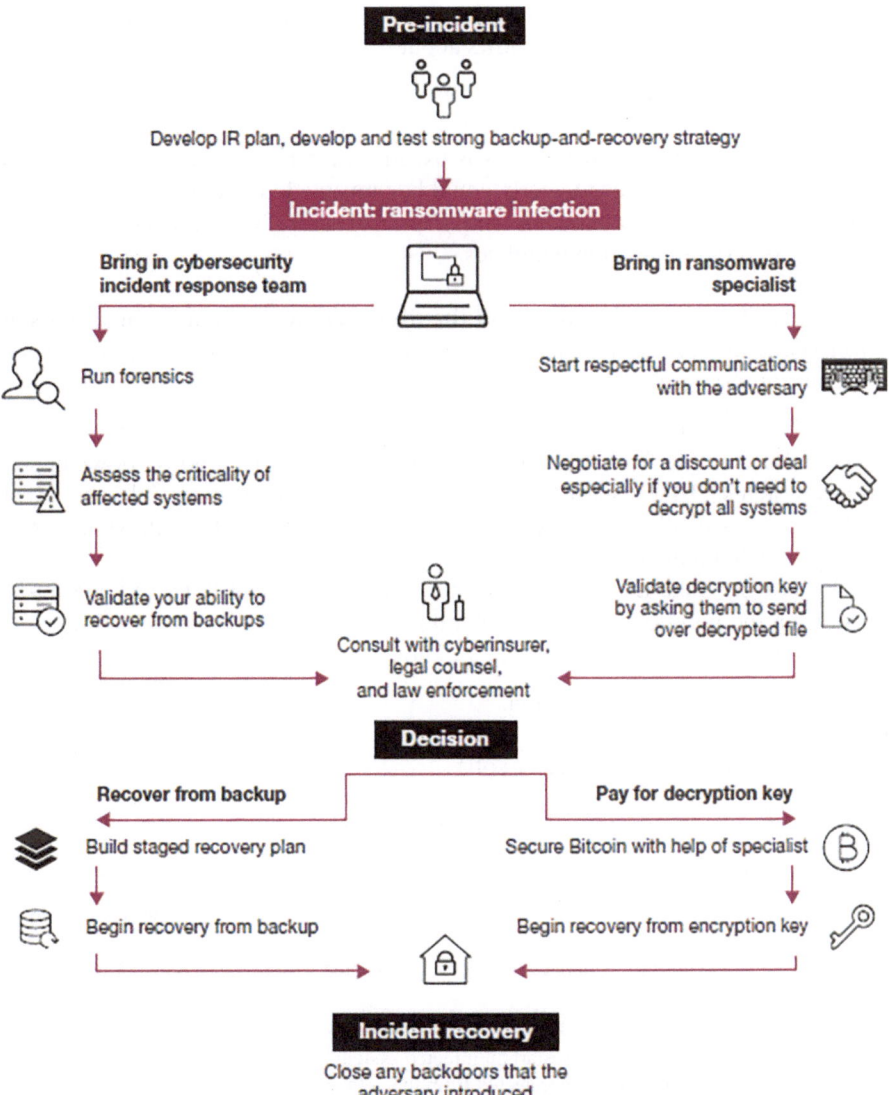

Fig. 4.1 Ransomware incident response process. (Source: Forrester Research (Zelonis 2019) via (Dignan 2019))

the facts. Take the actions you are supposed to take. Be prepared to answer these questions:

- What is the issue?
- How do you know there is an issue?
- What is the impact of the issue?
- Who is experiencing the issue?
- What have you done up to this point?

Anyone you speak with will want to know what happened, how you know it happened, the impact, and who else knows about it.

Do Not Connect External Drives or Offline Backups to the Network Companies and individuals have had their backup or offline data encrypted because they reconnected it to an infected network. This is usually caused by failure to recognize the threat. Connecting devices too early could be prevented via timely communication and training. CISA recommends the *Public Power Cyber Incident Response Playbook* for communication templates.[51]

The Great Shut Down, Reboot, Pull the Plug Debate To shut down or not shut down a computer during a ransomware incident is one of the most debated topics in ransomware response.

Table 4.1 contains advice from the major entities that discuss the topic.

The advice ranges from "shut down if you must," to "do not power down or shut off any systems." IBM's and EMSISOFT's advice are the most sensible, though all of the advice is challenging to execute during an incident. The following section examines each unique piece of advice.

CISA says, "Take the network offline at the switch level." The advice makes sense from the point of view of not shutting down systems but is very difficult to execute on a complex network. There are three ways to accomplish this.

- Change the switch configuration.
- Pull cables out of ports
- Power down the switch (hard or soft)

Table 4.1 Advice on whether to restart an infected computer during a ransomware incident

CISA	"Take Network offline at switch level." If not possible, "power down to avoid further spread."
EMSISOFT	"Do not restart impacted devices." "impacted systems should be put into hibernation."
FTC	"Immediately disconnect the infected computers or devices from the network."
IBM	"Do not reboot or restart an infected system." "the infected system should be hibernated and disconnected from the network immediately." "shut down immediately" if encryption has begun but not completed (no ransom note displayed).
Justice Department	"Remove from the network as soon as possible." "isolate or power-off affected devices that have not yet been completely corrupted."
Secret service	"Do not power down or shutoff any systems affected by ransomware."

---

[51] American Public Power Association (2019).

Changing the configuration requires a network engineer with a clean, uninfected computer or local access. Pulling this together quickly is difficult, even with practice and complete disregard for the consequences.

Pulling cables out of ports can work if the goal is to isolate the switch from the rest of the network, but it is not the best way to isolate computers from each other. It is also likely that cables would be disconnected carelessly, making reconnection problematic.

Powering down the switch is effective but requires either a network engineer or someone with physical access at the location. Additionally, configurations are frequently not saved permanently, meaning that the configuration could be lost when the switch is powered down.

One other consideration with action at the switch level is that computers often have both wired and wireless connections. The computer will switch to a wireless connection if the wired connection is not available. It is possible to power down a switch and not realize that computers are still communicating on a separate dedicated wireless network.

All of the options above are considered preferable to the impacts of a full-blown ransomware attack, but none are sufficient to stop the outbreak quickly without collateral damage.

The FTC, IBM, and DOJ recommend disconnecting devices from the network. This is good advice for small outbreaks. IT can instruct every user to disconnect the network cable, which is relatively easy. Responders should remember that computers might also be connected to WiFi. IBM's advice is a slightly better in this case because they suggest hibernating the machine.

EMSISOFT and IBM suggest hibernating affected machines. This is probably the best advice if the organization can prove through testing that hibernation works. An even better option is forced hibernation via remote commands. That way, the responder can have some assurance that it worked.

Shutting down infected machines is the go-to move for many responders, especially when it comes to ransomware. Most of the advice recommends not doing that because it can harm an investigation, but good luck convincing an IT manager to not power down in the heat of the moment. This is an area where IBM provides conflicting advice. Hibernate if the ransom note is displayed. Shut down if it is not. Powering down or rebooting can have negative consequences for recovery. One example is the Jigsaw ransomware. Jigsaw deletes files if a victim does not purchase a decryption key in time.[52] Rebooting the machine hastens the deadline.[53] Research from Michigan State University published by the IEEE demonstrated that it is possible to recover encryption keys from the memory of infected systems when the encryption is simple or improperly configured.[54]

---

[52] ENISA (October 20, 2020).

[53] Ibid.

[54] Baipal and Enbody (2020).

The shutdown vs. no shutdown debate is a difficult one for organizations. They should ask the following questions as they devise their Incident Response plans:

- Is the advice the same for one infected machine vs. multiple infected machines?
- Do we shut down devices that have not displayed a ransom note?
- Is the advice the same for organizations with very sophisticated cybersecurity teams vs. those with limited teams?
- Is it more important to preserve evidence or data? Is there a balance?

The best option for organizations is some form of "kill switch" that disables network access at the device, VLAN, or segment level. They need a way to cut off network access to most network connected devices instantly. Reconfiguring switches and pulling cables out of machines is too time consuming and can be risky. Software installed at the endpoint level or switching level can be used to instantly cut network access for a period of time. There are at least four solutions with this functionality.

- McAfee Network Security Platform: Quarantine Hosts. It works based on "sensor" information or manual action.[55]
- FireEye HX: "Isolate compromised devices with a single click."[56]
- Airgap Networks: Ransomware Kill Switch. A company named Airgap Networks astutely noticed the kill switch need and created a product around it. The product falls short of the full requirement at the time of this writing because it only kills lateral traffic and not internet access.[57]
- Darktrace Antigena/RESPOND: This tool detects anomalous network activity and can automatically disrupt it if configured to do so.[58]

There are potential disadvantages to relying on tools such as these.

1. Ransomware frequently attacks security products and tries to disable them.[59] If the security product is disabled, then responders will have to return to manual methods.
2. Kill switch functionality is powerful. It would have to reside behind several security layers (MFA, Dual Key, among others).
3. The kill switch could be attacked and disabled. Attacking the kill switch system should be threat modeled. How can an astute adversary disable the kill switch during an attack?
4. How is the "kill" command reversed? McAfee has had kill switch functionality for many years. A McAfee administrator at one company accidentally invoked it, not realizing that the invocation disabled all network communication to impacted endpoints and required manually deleting a text file on each computer restore

---

[55] McAfee (2020).

[56] FireEye (2016).

[57] Airgap Networks (2020).

[58] Darktrace (2021).

[59] Cimpanu, Ransomware installs Gigabyte driver to kill antivirus products (2020d).

network access. Administrators must fully understand how the kill switch works and how to reverse it.

Communication Communication is vitally important in all stages of an incident, especially early on. Someone must be responsible for developing messages for employees, customers, and management. In particular, employees will inform friends, family, and social media of a "hacking" or ransomware incident unless instructed otherwise. A recent example of this is the September 2020 UHS ransomware incident. Employees used Reddit and other social media platforms to inform the world of a ransomware outbreak before the company made a public announcement.[60] This is not totally unexpected in a company with hundreds of sites and thousands of users. Organizations should be aware that "inside" information will reach the public relatively quickly. Security company Huntress recommends, "Don't answer the phone until legal gives you advice."[61] The reason is that communication during an incident is discoverable in subsequent litigation.[62] This does not mean that victims should go quiet. It means that they should be cautious. They should seek legal advice, send meaningful communication, and stick to the plan.

Out of Band Communication Organizations should have an "out of band" communication system that incident response teams can rapidly deploy in case email and corporate chat services are down. In many cases, inaccessible systems force organizations to use SMS apps in a "call tree" fashion because internal systems are inaccessible.[63] The out of band system should remain secret until it is needed, so it is not easily discoverable by attackers.

Expect the Unexpected Ransomware incidents can have a sci-fi level of unpredictability. The criminals who deploy ransomware have a single goal, which is financial profit. They will use almost any means possible to maximize pressure, leading to payment. There are many real-world examples.

- Ransom demand sent directly to the insurance company[64]
- Printers and Point of Sale devices printing Egregor ransom notes[65]
- Ransomware operator contacts a reporter ahead of victim interview with the same reporter[66]
- Ragnar Locker Team runs Facebook ads pressuring Campari Group to pay[67]

---

[60] Cimpanu, UHS hospital network hit by ransomware attack (2020f).

[61] Hanslovan and Bisnett, hack_it 2020 (2020).

[62] Freedman, Security Economy Episode 3: Got a Privacy Plan? GDPR, CCPA, and the Rise of Ransomware (2020b).

[63] Bracy (2020).

[64] Gibson and Laporte (2019).

[65] Abrams, Egregor ransomware print bombs printers with ransom notes (2020b).

[66] Krebs, The Hidden Cost of Ransomware: Wholesale Password Theft (2020b).

[67] Krebs, Ransomware Group Turns to Facebook Ads (2020a).

These are just a few examples. There are many more. Ransomware operations run like businesses, so maximizing profits and protecting reputations is important to operators. The fact that there are no rules for criminals means that all kinds of unexpected events are possible.

Response

Ransomware in Incident Response Plans One of the critical factors that may be catching organizations by surprise is that ransomware is not explicitly included in incident response plans. Most incidents that organizations face are relatively ordinary. A ransomware outbreak will test every part of the plan. This research included a survey for Information Security stakeholders. The survey asked two questions about Incident Response plans.

Does your organization have an incident Response Plan? (Table 3.4)

- 95.65% Yes
- 4.35% No

It then asked, "Does your organization have an incident response plan for ransomware? (Table 3.5)

- 86.21% Yes
- 13.79% No

There is a 9% decrease in the number of organizations with incident response plans for ransomware. Overall, 15.94% of the respondents did not have an incident response plan or a ransomware plan. This is not just an issue for small companies. Fifty percent of the respondents from organizations with at least 1000 users said they did not have a plan for ransomware. (Table 3.6)

There are two ways of thinking about whether ransomware should be included in incident response plans.

1. Ransomware has multiple unique factors that are explicitly addressed in an incident response plan.
2. Ransomware is not considered "special" in a comprehensive plan.

The first view is more reasonable because most incidents do not require spending money to prevent data destruction and disclosure. The fact that there might be a financial outlay to criminals is by itself enough to be worthy of addressing in a plan. Who will make the decision whether to pay a ransom? A security manager? A CISO? What about a vCISO? It will most likely be the CEO, General Counsel, and CISO in consultation with the CFO and "Compliance."

Law Enforcement Wants to be Informed of an Incident Multiple government-authored ransomware guides state that victims should inform law enforcement about ransomware incidents. The FBI and OFAC also want to know if victims pay a ransom. (Information about regulations on ransom payments is provided in the Ransomware Laws section.) The following groups request notification about ransomware incidents:

- FBI
- Secret Service
- IC3
- CISA

The topic of whom to call was discussed during the first day of the 3rd CISA Cybersecurity Summit in September 2020. Jason Conboy from the Department of Homeland Security said "call anybody."[68] He went on to say that the agency regionally closest to the victim is likely to respond. This comment should be tested before an incident.

Recovery

The response portion of a ransomware incident is easier than that of the recovery part. The recovery topics are far more complex, leading to the question of whether a ransom should be paid or even if a company can remain in business. The guides linked at the top of this section have plenty of advice on ransomware recovery. The section below discusses real-world scenarios and measures that organizations should consider.

Backup Encrypted Files Before Trying to Recover There have been cases where decryption tools accidentally damage files.[69] EMSISOFT and Dr. Johannes Ullrich from the SANS Institute recommend backing up files before attempting to decrypt them, so that victims can obtain a second chance to decrypt files if there is an issue.[70] EMSISOFT made this recommendation after a purchased Ryuk ransomware decryptor damaged files due to a bug in the decryptor.

Archive Files in Case There Is a Decryptor in the Future This suggestion also came from EMSISOFT. There are cases where ransomware operators collect ransoms with no intention to decrypt files. EMSISOFT recommends backing up files "in case a decryptor becomes available in the future, which has happened multiple times including the Hive ransomware takedown by the FBI in January 2023."[71, 72] Decryptors generally come from three places:

---

[68] Conboy (2020).

[69] EMSISOFT (2019).

[70] Ulrich (2019).

[71] EMSISOFT (n.d.).

[72] US Department of Justice (2023).

- "Official" decryptors purchased from attackers
- Cybersecurity industry
- Via the release of encryption keys by ransomware operators.[73]

Beware of Fake Decryptors (and Double Encryption) Opportunistic criminals prey on ransomware victims' lack of knowledge with enticing fake decryptors promising to restore access to encrypted files. The fake decryptors install additional ransomware, leading to a double ransomware situation. One of the ransomware strains preying on victims in this way is called STOP Djvu. It was the most commonly submitted ransomware on the ID Ransomware service (https://id-ransomware. malwarehunterteam.com) as of June 2020.[74] It should serve as a reminder to enterprise administrators that they should seek professional incident response assistance and that there are no shortcuts.

If You're Vulnerable to One Ransomware Attack, You're Vulnerable to a Second Restoring systems is only one aspect of ransomware recovery. The other is eliminating the vulnerabilities that allowed the first attack to succeed. It is not uncommon for organizations to experience multiple ransomware incidents, sometimes in close succession, likely reusing the initial unfixed vulnerability that caused the first incident. The Australia-based Toll Group was the victim of Netwalker in February 2020 and Netfilm in May 2020.[75] The National Veterinary Associates (NVA) was hit with Ryuk twice within the same calendar year.[76] Victims must take action to disinfect the network, reduce vulnerabilities, and improve security.

Betterment Betterment is an insurance concept in which an entity making a claim improves its situation with funds received from the claim. A simple example of this is filing a claim because a storm damaged the roof of a house. Using insurance funds to patch the roof is possible. Replacing the roof with a new, stronger one improves the protection offered by the house. Improving the status of a situation through insurance funds is known as "betterment." In terms of cybersecurity, this means that organizations improve their security posture after the claim. There are typically contractual arrangements over what falls under "betterment" and what percentage of the funds can fall into this category.[77] Simply restoring systems to their previous vulnerable state is unwise, and the remediation process should factor in improving security where reasonable. Jack Kudale discussed this topic in detail on the Cybersecurity Interviews Podcast with Douglas Brush.[78]

---

[73] Cimpanu, GandCrab ransomware operation says it's shutting down (2019a).

[74] Abrams, Fake ransomware decryptor double-encrypts desperate victims' files (2020c).

[75] Osborne (2020).

[76] Krebs, Ransomware Bites 400 Veterinary Hospitals (2019).

[77] Law Insider (n.d.).

[78] Kudale (2020).

Secure Cloud Backups Ransomware attackers will attack backups in two different ways. They will exfiltrate backups without victims noticing since company monitoring tools cannot see this, and they will delete them.[79] Organizations should work with cloud backup providers to determine what happens if backups are modified or deleted. Are they restorable? For how long? Any answer in which organizations permanently lose backups when deleted is unsatisfactory. This activity can begin before an incident as part of the preparation process.

Maintain a Software Library with Activation Keys One of the challenges of the recovery process is reinstalling operating systems and software. Software libraries will be attacked just like other data, and installation media is not always well maintained.[80] Organizations should maintain what ITIL calls a "Definitive Media Library," which contains all installation software, license information, and recovery keys. The library is useful on a daily basis, but it is critical to consider how it would be accessible in a disaster response scenario.

The Costs and Complexity of Recovering Data The costs and complexity of recovering data without paying a ransom are listed as two of the greatest motivators for paying ransoms. (Table 3.8) (Motivators are covered in the "Should You Pay" section.) Sometimes paying is not an option because decryptors fail, the ransomware is actually a wiper/locker, legal issues (sanctioned groups), or victims cannot or will not pay. Recovering from ransomware or other catastrophic data loss incidents is expensive and time-consuming, leading to paralyzed operations, lost revenue, and lost customer trust.[81] Catastrophic loss of data is not new. The means have changed from natural disasters and fires to cyberattacks. The 1974 US National Personnel Records Center fire provides lessons about recovery that are informative to digital disaster recovery.[82]

- Building space required for restored records. Is there adequate rack, power, and cooling if infected servers are replaced because investigators hold them as evidence or are out of date?
- How will staff roles and responsibilities change during the incident and the subsequent recovery? Ransomware incidents are "all hands on deck" events. A lot of the work is nontechnical (communication, records retrieval, purchasing, and customer response, among others). Staff burnout is a concern during any major event, and especially during a potentially catastrophic data loss incident.
- It was essential to keep people working without a loss of pay. Payroll delays and furloughs may result in a loss of staff that compounds the original event's damage
- A "lessons learned" process took place with the goal of achieving "total fire safety." Documenting lessons learned is critical for improving security and recovery procedures.

---

[79] Abrams, Ransomware Attackers Use Your Cloud Backups Against You (2020d).

[80] Johnson (2019).

[81] Abrams, Tyler Technologies paid ransomware gang for decryption key (2020f).

[82] Stender and Walker, The National Personnel Records Center Fire: A Study in Disaster (1974a).

Chapter 5
Proposed Solutions

Abstract This chapter examines the various approaches to mitigating the ransomware menace, emphasizing the need for global cooperation and innovative security methodologies. The profitability and ease of distribution of ransomware pose significant challenges, prompting the exploration of six potential solutions ranked by their feasibility and effectiveness. Establishing international norms and enhanced cooperation, including extradition agreements, stands out as a primary strategy, albeit contingent upon broad international agreements, which remains elusive due to varying national interests. Hardware solutions offer a technical countermeasure by detecting abnormal encryption activities, while adopting zero-trust architectures promises to reduce network vulnerabilities by enforcing strict access controls and segmentation. Additionally, reducing the profitability of ransomware through improved cybersecurity practices and considering the controversial steps of banning cryptocurrencies and ransom payments are discussed. The latter is likened to measures against kidnapping and maritime piracy and could decisively undercut ransomware operations but faces ethical, legal, and practical hurdles. This chapter underscores the complexity of combating ransomware and advocates for a comprehensive and collaborative approach that balances immediate protective measures with long-term deterrent strategies.

International Norms and Cooperation Norms are not laws. They are "principles that represent" the "collective expectations" of societies "and are both widely accepted and internalized by the members of the community within which they evolve."[1] The international community can severely reduce the threat of ransomware if enough countries agree to investigate and prosecute cases in a collaborative way. Extradition is particularly important so that law enforcement can bring cybercriminals to justice. A recent example is the indictment and arrest of Canadian citizen Sebastien Vachon-Desjardins for his alleged involvement with Netwalker ransomware.[2]

---

[1] Erskine and Carr (2016).

[2] Corder and Bajak (2021).

Of course, norms and cooperation only work if enough countries agree to follow them and follow through. However, this is unlikely to happen, as ransomware has become an important income source to which some nations are willing to turn a blind eye when used internationally.[3, 4] Fifty-five nations signed onto the Budapest Convention on Cybercrime, which resulted in documented improvements in mutual assistance but two of the world's major cyber powers, Russia and China have not signed.[567]

Norms can work but are unlikely to do so based on deeply divergent attitudes toward cybercrime from different governments as of 2021.

Hardware Solutions Ransomware, like other advanced malware, attacks security tools and tries to disable them. It also generates a large amount of hard disk and cryptographic activity when encrypting systems. A hardware-based system that detects anomalous disk activity, especially using encryption, could either pause the attack or take another action that forces the system operator to respond. Computer users would only accept this if it kept false positives to a minimum and could not be used to launch denial of service attacks on the system. It would also take a long time for hardware solutions to reach the adoption levels required to stop ransomware. Southern Methodist University researchers developed a software-based solution that they claim detects zero-day ransomware by observing power surges on the system.[8] A hardware-based solution that malware cannot disable and that can detect "rogue encryption" has the potential to inhibit or even stop ransomware.[9]

Zero-Trust Zero-trust is one of the biggest buzzwords of the 2020s. Zero-trust architecture is a methodology that assumes that the network is always compromised, so each connected system must meet specific requirements (patch levels, presence of security tools, and other compliance factors) to allow the system to connect to company resources (cloud services, files, and systems).[10] Its most valuable attribute is built-in device segmentation. Zero-trust makes it difficult for attackers to move laterally across a network because client systems are not on the same network. Instead, they use authentication and encryption to access independent cloud-based services.[11]

The combination of ransomware, pandemic related "Work from Home," and the need for device segmentation drove zero-trust to the top of IT professionals' wish

---

[3] Schwartz (2020).

[4] Ilascu (2021).

[5] Chernenko et al. (2018).

[6] Council of Europe (2001).

[7] Council of Europe (2020).

[8] Taylor et al. (2020).

[9] Ibid.

[10] NIST Computer Security Resource Center (2020).

[11] Google (n.d.).

list. Zero-trust could play a vital role in solving the ransomware problem because there would not be much of a network for attackers to compromise. They could attack individual systems that connect to cloud-based services, but they could not move from system to system across a network. Cloud-based file services are the weak point in this scenario. Service providers and customers would need to secure their services from ransomware and would require the ability to restore to a pre-compromise state. OneDrive currently has that capability.

Zero-trust is worth the hype when it comes to achieving long-held best practices such as device segmentation, MFA, and encryption.

Reduce the Profitability of Ransomware The incident response company Coveware introduced the concept of "unit economics" to the ransomware discussion.[12] This means "profitability per unit."[13] In the ransomware game, it is the profitability per victim. It is how much profit remains after compromise-related expenses. One of the goals of cybersecurity is to make it time-consuming and expensive for criminals to perform attacks. This is how "defense in depth" is justified. The more security layers there are, the longer it will take to compromise a victim, giving them a chance to detect and stop an attack. Multiple layers of security may motivate attackers to move on to easier targets.

There are ways to increase costs for attackers. The first is to implement the basics. Many attacks are the result of phishing, exposed RDP, and unpatched VPNs. Implementing MFA, moving services inside the firewall (or to the cloud), and patching VPNs removes the most common and easy to exploit attack vectors. These best practices will slow attackers, making it more time-consuming and challenging to launch ransomware attacks. This solution will succeed if nearly all companies implement good cyber hygiene practices in their organizations.

Banning Cryptocurrencies Cryptocurrencies are one of the factors that make ransomware possible because of easy and (relatively) anonymous payments. Banning cryptocurrency would force attackers to use the much more traceable banking system. US-based financial institutions automatically file "Suspicious Activity Reports," alerting the government to the money transfer.

It remains unlikely that governments will universally ban cryptocurrencies, leaving plenty of well-financed victims to attack. India announced a possible ban on cryptocurrencies in 2021, but they may replace the "private" ones (bitcoin) with a government-sponsored version.[14] It is impossible to determine whether ransomware gangs will move to the "official" cryptocurrency or cease operations in India. This will be an intriguing experiment as India was the 2nd highest victim of ransomware by country in late 2020.[15] Will it become impossible for companies in India to pay?

---

[12] Coveware (2020).

[13] Lighter Capital (n.d.).

[14] Jha (2021).

[15] Business Insider (2020).

If so, will ransomware gangs cease Indian operations? This is not a ransom ban, so large organizations might be able to order payments through other countries where cryptocurrencies are legal. Nevertheless, a ban on cryptocurrencies potentially removes one of the three Cs of ransomware, significantly reducing the effectiveness of potential attacks.

Ban Ransom and Extortion Payments   Banning ransom and extortion payments as a form of deterrence is the solution most likely to defeat ransomware over the long term. This would cut off funds to ransomware actors, completely disincentiv-izing them from using ransomware as a tactic. Yvonne Dutton and Jon Bellish wrote a 2014 paper exploring the effects of banning ransom payments in maritime piracy cases. Piracy and ransomware pose similar problems, with the difference being human lives are at stake in most piracy cases. The takeaway from the report is that banning ransom payments would limit the options of victims, their employers, and their loved ones to return victims safely.[16] This opinion is shared by "maritime orga-nizations representing ship owners."[17] Lives are not usually directly at risk with ransomware. Instead, banning ransom payments impacts the viability of organiza-tions and the livelihoods of those that work for them. This poses completely differ-ent moral and ethical challenges. Banning ransom payments would force companies to admit that there was a breach (where legally applicable), admit to losing data, and potentially put individuals and other companies at risk when data is published, pos-sibly resulting in the breached company going out of business. This can lead to the loss of jobs and even potential legal liability for victims. It has not been acceptable up to this point to risk the lives of hostages to achieve the long-term goal of ending piracy. Would society and the corporate world accept the destruction of businesses to achieve the long-term goal of ending ransomware? There is a difference between paying ransom to protect lives and paying it for data.

The effects of banning ransoms have not been tested. However, Italian law requires that prosecutors approve ransom negotiations and payments for kidnapping victims in consultation with the police.[18] In the seven years since the law passed, the number of kidnappings declined from fifty to sixty per year to approximately five.[19] Dutton and Bellish suggested that this model might reduce maritime piracy by giv-ing law enforcement the ability to track pirates.[20] Unfortunately, this is unlikely to work in the digital world where criminals never have to expose themselves to law enforcement and foreign militaries.

Banning ransom payments is the most consequential option. It is the most likely to work but will also cause the most short-term damage. It requires a long-term commitment and international cooperation to become a viable solution to the

---

[16] Dutton and Bellish (2014).

[17] Ibid.

[18] Bohlen (1998).

[19] Gumbel (2011).

[20] (Dutton and Bellish 2014)

problem, understanding that harm is inevitable in the short term. Lawmakers need to decide if the short-term harm to individuals is worth a long-term solution to the problem.

Additional reading on the topic of banning ransom payments for piracy is found here: Ship-Owners and the Twenty-First Century Somali Pirate: The Business Ethics of Ransom Payment. (https://www.jstor.org/stable/41476315)

Appendixes

Appendix A: Latest Internet Attack Holds Computer Files Hostage

Latest Internet attack holds computer files hostage

Tuesday, May 24, 2005 Updated at 6:24 AM EDT

Associated Press

Washington — Computer users already anxious about viruses and identity theft have new reason to worry: Hackers have found a way to lock up electronic documents and then demand money to get them back.

Security researchers at San Diego-based Websense Inc. uncovered the unusual extortion plot when a corporate customer they would not identify fell victim to the infection, which encrypted files that included documents, photographs and spreadsheets.

A ransom note left behind included an e-mail address, and the attacker later demanded $200 (U.S.) for the digital keys to unlock the files.

"This is equivalent to someone coming into your home, putting your valuables in a safe and not telling you the combination," said Oliver Friedrichs, a security manager for Symantec Corp.

The FBI said the scheme, which appears isolated, was unlike other Internet extortion crimes. Leading security and antivirus firms this week were updating protective software for companies and consumers to guard against this type of attack, which experts dubbed "ransom-ware."

"This seems fully malicious," said Joe Stewart, a researcher at Chicago-based Lurhq Corp. who studied the attack software.

Mr. Stewart managed to unlock the infected computer files without paying the extortion, but he worries that improved versions might be more difficult to overcome.

Internet attacks commonly become more effective as they evolve over time as hackers learn to avoid the mistakes of earlier infections.

H. Halikias, *Digital Shakedown*, https://doi.org/10.1007/978-3-031-65438-1

"You would have to pay the guy, or law enforcement would have to get his key to unencrypt the files," Mr. Stewart said.

The latest danger adds to the risks facing beleaguered Internet users, who must increasingly deal with categories of threats that include spyware, viruses, worms, phishing e-mail fraud and denial of service attacks.

FBI spokesman Paul Bresson said more familiar Internet extortion schemes involve hackers demanding tens of thousands of dollars and threatening to attack commercial websites, interfering with sales or stealing customer data.

Experts said there were no widespread reports the new threat was spreading, and the website was already shut down where the infection originally spread. They also said the hacker's demand for payment might be his weakness, since bank transactions can be traced easily.

"The problem is getting away with it -- you've got to send the money somewhere," Mr. Stewart said. "If it involves some sort of monetary transaction, it's far easier to trace than an e-mail account."

Sources: (Associated Press 2005) https://seclists.org/interesting-people/2005/May/249

(Associated Press 2005) https://advance-lexis-com.revproxy.brown.edu/api/document?collection=news&id=urn:contentItem:4NGB-XWG0-TXJC-4107-00000-00&context=1516831

Appendix B: Technology of Ransomware: Additional Resources

Ransomware is not a Windows-only problem. Here are some additional resources that cover other technologies:

- EvilQuest Mac Ransomware: https://objective-see.com/blog/blog_0x59.html
- Android Ransomware: http://www.welivesecurity.com/2016/02/18/the-rise-of-android-ransomware
- Ragnar Locker used an old version of Virtual Box and a Windows XP VM to execute its payload: https://news.sophos.com/en-us/2020/05/21/ragnar-locker-ransomware-deploys-virtual-machine-to-dodge-security/
- Attempt to use Zero Day on Sophos Firewall OS to launch ransomware attack: https://news.sophos.com/en-us/2020/05/21/asnarok2
- Snatch ransomware reboots Windows into Safe Mode to bypass anti-virus: https://news.sophos.com/en-us/2019/12/09/snatch-ransomware-reboots-pcs-into-safe-mode-to-bypass-protection/
- Linux Ransomware: https://www.bleepingcomputer.com/news/security/ransomexx-ransomware-also-encrypts-linux-systems/
- Ransomware written in .Net: https://www.exploit-db.com/docs/47680

- Databases encrypted, companies extorted, in database ransomware campaigns: https://www.bleepingcomputer.com/news/security/hacker-extorts-online-shops-sells-databases-if-ransom-not-paid/
- EKANS ransomware targeting ICS systems: https://www.fortinet.com/blog/threat-research/ekans-ransomware-targeting-ot-ics-systems
- Bitlocker based ransomware: https://www.blackhillsinfosec.com/bitlocker-ransomware-using-bitlocker-for-nefarious-reasons
- Ryuk used blockchain based DNS in attacks: https://www.grc.com/sn/sn-789-notes.pdf

One of the best sources available for ransomware and other attack profiles is https://thedfirreport.com

Appendix C: Cybersecurity Guides for Small Businesses

- Cyber guides from the US Government: https://www.fcc.gov/general/cybersecurity-small-business
- Small Business Administration: https://www.sba.gov/business-guide/manage-your-business/stay-safe-cybersecurity-threats
- FTC: https://www.ftc.gov/tips-advice/business-center/small-businesses/cybersecurity
- NY State: https://www.nyssbdc.org/resources/cybersecurity.html
- ENISA: https://www.enisa.europa.eu/publications/ransomware
- Australia: https://www.cyber.gov.au/acsc/view-all-content/essential-eight

Appendix D: Healthcare Industry Resources

- 2020 CISA/FBI/HHS Joint Cybersecurity Advisory: Ransomware Activity Targeting the Healthcare and Public Sector:
- https://us-cert.cisa.gov/sites/default/files/publications/AA20-302A_Ransomware%20_Activity_Targeting_the_Healthcare_and_Public_Health_Sector.pdf
- 2020 CISA, MS-ISAC Ransomware Guide:
- https://www.cisa.gov/sites/default/files/publications/CISA_MS-ISAC_Ransomware%20Guide_S508C.pdf
- HHS FACT SHEET Ransomware and HIPAA:
- https://www.hhs.gov/sites/default/files/RansomwareFactSheet.pdf
- HHS HC3 Products. This is a great source of TTP information that is useful across sectors
- https://www.hhs.gov/about/agencies/asa/ocio/hc3/products/index.html#sector-alerts
- HICP: Health Industry Cybersecurity Practices

- https://www.phe.gov/Preparedness/planning/405d/Pages/hic-practices.aspx
- H-ISAC: Health ISAC
- https://h-isac.org
- CIS: No-Cost Malicious Domain Blocking and Reporting for U.S. Hospitals
- https://www.cisecurity.org/hospitals/

Appendix E: 2020 State Legislation Related to Ransomware or Computer Extortion

NATIONAL CONFERENCE OF STATE LEGISLATURES
7700 EAST FIRST PLACE   DENVER, COLORADO 80230
303-364-7700   FAX: 303-364-7800

2020 State Legislation Related to Ransomware or Computer Extortion
As of Sept 23, 2020

Summary: Measures were introduced in at least nine states in 2020. Indiana and Virginia adopted measures to provide for studies and West Virginia enacted legislation enhancing criminal penalties for ransomware.

Bill	Status	Summary
Calif. S.B. 922	Failed - Adjourned	Requires the prosecution for a felony violation of specified computer related crimes, including introducing ransomware into a computer with intent to extort property from another, to be commenced within three years after discovery of the commission of the offense. Requires the filing of a criminal complaint within nine years of the commission of the offense.
Conn. H.B. 5511	Failed - Adjourned	Requires that the Commissioner of Emergency Services and Public Protection analyze municipal cybersecurity needs throughout the state and determine the feasibility of the Department of Emergency Services and Public Protection providing individualized assistance to municipalities most at risk of suffering cybersecurity attacks.
Georgia H.B. 1133	Failed - Adjourned	Relates to general provisions of state government so as to prohibit state agencies from paying ransoms in response to cyber attacks, provides for a definition, provides for related matters, provides for an effective date, repeals conflicting laws.
Iowa S.B. 2080	Failed - Adjourned	Prohibits the state and political subdivisions of the state from expending public moneys for payment to persons responsible for ransomware attacks.
Iowa S.B. 2391	Failed - Adjourned	Prohibits the state and political subdivisions of the state from expending public moneys for payment to persons responsible for ransomware attacks.
Indiana SR 13	Adopted	Relates to legislative council to assign the topic of the potential dangers of cyber hacking in state government specifically the use of ransomware, urges the legislative council to assign to an appropriate study committee the topic of the potential dangers of cyber-hacking in state government, specifically the use of ransomware.
Maryland H.B. 215	Failed - Adjourned	Prohibits a person from knowingly possessing certain ransomware with the intent to use that ransomware for introduction into the computer, computer network, or computer system of another person without the authorization of the other person, establishes that a person who violated the Act is guilty of a misdemeanor and on conviction is subject

Bill	Status	Summary
		to imprisonment not exceeding 10 years or a fine not exceeding ,000 or both, applies the Act prospectively.
Maryland H.B. 635	Failed - Adjourned	Prohibits a person from knowingly possessing certain malware or ransomware with the intent to use that malware or ransomware for the purpose of introduction into a computer, computer network, or computer system of another person without the authorization of the other person, creates a certain exception, establishes a certain penalty.
Maryland S.B. 30	Failed - Adjourned	Relates to crimes involving computers.
New York S.B. 7246	Pending	Creates a cyber security enhancement fund to be used for the purpose of upgrading cyber security in local governments, including but not limited to, villages, towns and cities with a population of one million or less and restricts the use of taxpayer moneys in paying ransoms in response to ransomware attacks.
New York S.B. 7289	Pending	Prohibits any municipal corporation or other government entity from paying ransom in the event of a cyber-attack against such municipal corporation's or government entity's critical infrastructure.
Virginia HJR 64	Adopted	Requests the Information Technologies Agency to study the Commonwealth's susceptibility, preparedness, and ability to respond to ransomware attacks, provides that in conducting its study, the Agency shall assess the Commonwealth's susceptibility to ransomware attacks at the state and local levels of government.
West Va. S.B. 261	Enacted	Creates criminal penalties for introducing ransomware into any computer, computer system, or computer network with the intent to extort money or other consideration, sets forth the elements of the offense, establishes criminal penalties.

NCSL contact for additional information: Pam Greenberg, NCSL Denver Office, pam.greenberg@ncsl.org

Powered by
LexisNexis® State Net®

LexisNexis Terms and Conditions

Source: Pam Greenberg, NCSL Denver Office. Cannot locate online as of 3/1/2021 (The United Nations maintains a list of global cybercrime laws: https://sherloc. unodc.org/cld/v3/cybrepo/)

Appendix F: NY State Senate Bill 7246

Sponsored by Senator Phil Boyle (R) 4th Senate District

https://www.nysenate.gov/legislation/bills/2019/s7246

```
S7246 Bill Text
                    S T A T E     O F     N E W     Y O R K
_____

_____


                                  7246

                            I N   S E N A T E

                          January 14, 2020
                         _____
```

Introduced by Sen. BOYLE -- read twice and ordered printed, and when printed to be committed to the Committee on Veterans, Homeland Security and Military Affairs

AN ACT to amend the executive law and the state finance law, in relation to cyber security enhancement funding; and to restrict the use of taxpayer moneys in paying ransoms

THE PEOPLE OF THE STATE OF NEW YORK, REPRESENTED IN SENATE AND ASSEMBLY, DO ENACT AS FOLLOWS:

Section 1. The executive law is amended by adding a new section 719 to read as follows:

§ 719. CYBER SECURITY ENHANCEMENT. 1. THE COMMISSIONER OF THE DIVISION OF HOMELAND SECURITY AND EMERGENCY SERVICES SHALL ESTABLISH A PROGRAM FOR THE PURPOSE OF UPGRADING CYBER SECURITY IN LOCAL GOVERNMENTS, INCLUDINGBUTNOTLIMITEDTO,VILLAGES,TOWNSANDCITIESWITHAPOPULATION OF ONE MILLION OR LESS.

2. ALL COSTS OF SUCH PROGRAM SHALL BE PAID FOR BY THE CYBER SECURITY ENHANCEMENT FUND, ESTABLISHED BY SECTION NINETY-ONE-H OF THE STATE FINANCE LAW.

§ 2. The state finance law is amended by adding a new section 91-h to read as follows:

§91-H.CYBERSECURITYENHANCEMENTFUND.1.THEREISHEREBYESTABLISHED IN THE JOINT CUSTODY OF THE STATE COMPTROLLER AND THE COMMISSIONER OF TAXATION AND FINANCE A FUND TO BE KNOWN AS THE CYBER SECURITY ENHANCEMENT FUND.

2. SUCH FUND SHALL CONSIST OF FIVE MILLION DOLLARS TRANSFERRED BY THE STATE COMPTROLLER FROM THE GENERAL FUND, AND CONTRIBUTIONS CONSISTING OF GRANTS, INCLUDING GRANTS OR OTHER FINANCIAL ASSISTANCE FROM ANY AGENCY OF GOVERNMENT AND ALL MONEYS REQUIRED BY THE PROVISIONS OF THIS SECTION OR ANY OTHER LAW TO BE PAID INTO OR CREDITED TO THE FUND.

3. THE MONEYS FROM THE CYBER SECURITY ENHANCEMENT FUND SHALL BE DISBURSED BY THE DIVISION OF HOMELAND SECURITY AND EMERGENCY SERVICES TO BEEXPENDEDTOUPGRADETHECYBERSECURITYOFLOCALGOVERNMENTS,INCLUD-

ING BUT NOT LIMITED TO, VILLAGES, TOWNS AND CITIES WITH A
POPULATION OF
EXPLANATION--Matter in ITALICS (underscored) is new; matter in brackets
[ ] is old law
to be omitted.

LBD14014-03-0

S. 7246   2

ONE MILLION OR LESS, PURSUANT TO SECTION SEVEN HUNDRED
NINETEEN OF THE
EXECUTIVE LAW.
§ 3. Notwithstanding any other provision of law, after January first,
two thousand twenty-two, local and state taxpayer moneys shall not be
used to pay ransoms in response to ransomware attacks.
§ 4. This act shall take effect immediately.

Appendix G: Interview with NY State Senator Phil Boyle

10/1/2020
Interview with NY State Senator Phil Boyle – Video Call – Transcript – Updated for
  readability.

Harry Halikias My name is Harry. I'm a graduate student at Brown University
working on a Ransomware project, and I saw that you proposed a bill last year. I
believe was last year or early this year that would outlaw the payment of ransoms.
So I have a bunch of questions for you if that would be okay.

Philip Boyle Sure.

Harry Halikias All right, your, your legislation, where I live, they use the state or
local taxpayer funds from being used to pay ransoms. Why did you propose this
legislation?

Philip Boyle Well, I did so after numerous news reports of different municipal
government's, city, county, school districts, hospitals, municipal hospitals paying
ransom for cyber-attacks and as a fiscal conservative, it was driving me crazy that
taxpayer money was going towards paying these ransoms.

What my legislation proposes that the municipalities have a certain period of
time to upgrade (it wasn't like it starts today) have a couple years to upgrade their it
to protect against these cyber attacks. But there's a date certain after which no taxpayer money.

Harry Halikias I did notice the cyber security enhancement fund that you added to your bill, which others have not proposed.

What would the fund do and how did you come up with that number?

Philip Boyle That's a good question. So I actually attended a cybersecurity of forum, I think, was seems like last year, I guess it was earlier this year pre-pandemic arm and they talked about there was cybersecurity experts, even from the FBI and law enforcement and other and private companies.

And I asked them, or my staff, ask them. What do you think would be needed to upgrade? You know, and so they and I asked them for two things. I asked him to the approximate amount of money they thought should go into a fund like this, and the approximate amount of time that we needed to upgrade the systems. And they basically came upon two years from that date. You know, so whenever the legislation passed. And so, that's where the number comes from. It's nothing calculated. It's a general idea if we have to put more in will put more in. That's fine.

Harry Halikias Okay, thank you. Uh, what do you expect to change if your bill becomes law?

Philip Boyle I think it's going to open a discussion, because I think a lot constituents, voters, citizens, don't know about this. I look at cases like Atlanta and Baltimore, where they paid a million dollars or whatever it was, and it's now it's affecting school districts. And as you know, coming from Long Island, the school taxes here are very high, and we have very good schools. But our homeowners and property taxpayers do not want to see that their tax money is going to some 17-year-old criminal somewhere in Eastern Europe who decided to do a cyber-attack on their school district, and they got caught.

I think they we have to give an opportunity to upgrade their cyber security systems, but at some point in time, the hammer needs to fall.

Harry Halikias Does your bill prohibit the use of insurance funds to pay ransoms?

Philip Boyle No, it does not know. I mean, if that's the case, and obviously that's a growing industry. If that's what they want to do, put some towards insurance. I think it would be part of the overall insurance for the school district or the town or village, that's fine, and maybe a little more of a premium. I really have no problem with that.

Harry Halikias One of the things going on right now with the insurance industry is, because they've been forced to pay all this money, they're now scrutinizing the cybersecurity of the groups that they're insuring much more closely.

Philip Boyle Interesting

Harry Halikias And so they're looking at them and they're saying, you got to turn this on and turn this off.

Philip Boyle I believe that. That's very good.

Harry Halikias Yeah, it's good progress. A recent development of ransomware is that groups are stealing data and then demanding a payment to destroy the data and provide a certificate of destruction.

Experts argue that it might be better to pay a ransom to have data destroyed than to have it released to the public. What are your thoughts on this?

Philip Boyle Well, what kind of data we talking about? Something that's embarrassing to the company or the government or is it proprietary, you mean?

Harry Halikias That's a great question. So it depends on the organization. So I worked for a technology company. In my case it would be company data, potentially sensitive non-public information, trade information. Well, when it comes to municipalities, it could be information about citizens, maybe tax payments, grievances, board records, things like that. So it just depends on the organization.

Philip Boyle Yeah, that's interesting. I think that would be up to the elected officials in that municipality, whether it's a school board or the or the mayor or whatever the Village Board.

My concern with that is when you get a certificate of destruction, what does that really mean? We're talking about someone who did a crime to get to this point. Now you're going to take their word. Oh yeah, we have you pay us half a million dollars. We promise we're going to destroy it. You know, so it's a little questionable there.

Harry Halikias Yeah, and it's a big debate right now. It's a lot for us to think about because a year ago when this first came up or when we were first started talking about this, I think you and me separately, that wasn't a factor. Back then the advice was, well just restore your data and you go on with your life. Now, there's this other angle and it's controversial inside companies and inside of governments.

Philip Boyle That's a great point. This actually might also be the way that they, the criminals or the cyber terrorists if you will, that they're probably figuring out that the next thing. Because a lot of these municipalities and school districts are getting better at, "Okay. We have to back everything up in a separate location, so God forbid something like this happens, we can say no we're not paying you a dime." But if it's embarrassing information, then that's something a whole new area of crime, shall we say.

Harry Halikias They're finding new ways to squeeze everybody, you know make them pay.

Do you feel that the loss of data, well this kind of goes to the way we used to think about Ransomware where if you didn't pay you might lose your data, and that's the end of it. Do you feel that the loss of data might lead to a loss of trust in government?

Philip Boyle Absolutely no question about it you know and with that as we're talking about that if you have years and years, decades really of files and information that are suddenly gone, I think about it in a situation. So well before you were born, there was a big fire in the Veterans Administration in Washington DC and countless records were lost. World War Two veterans and all their military records, you know. And so since then, I think there's been a lack of, "oh, did you really serve" or, you know, "where's your official paperwork, it was destroyed in the fire." You know, this would be a cyber version of that. You know, there is definitely a loss of trust.

Harry Halikias I wasn't aware of that.

Philip Boyle Yeah.

Harry Halikias Municipalities in Connecticut, Rhode Island Massachusetts are getting hit hard by ransomware. Some are insured through local municipality trusts. How will the state help municipalities?

Philip Boyle I can see I would support the idea of New York, just like creating the fund; there would be also another way to financially assist there. Unfortunately, in this pandemic situation we don't have any money in Albany, or any other state, probably. But in normal times I would have no problem with the state lending hand and also some expertise to provide to these localities or municipalities to say this is what you need to do protect yourself. I mean a lot of them, they will contract with private companies, obviously, but states can help out to.

Harry Halikias All right. And the last question I have written down… According to a 2019 poll by Morning Consult, 56% of respondents oppose paying ransoms using taxpayer money, but 54% oppose increasing taxes to make cyber security improvements. What is your message to the public about the security of government resources?

Philip Boyle I would say that paying a little bit now for prevention is worth [it compared to] paying a lot down the road when the cyber ransom attack takes place. Because these ransom, they're not asking for a few dollars. They want millions and for a school district or a village, it could it could ruin them. That's what I would say.

Harry Halikias That's the same exact argument we've had in the corporate world so.

Philip Boyle I'm sure.

Harry Halikias Well, that's all the questions I had for you.

Philip Boyle Well, thank you. Good luck with your, your project.

Harry Halikias Thank you. I really appreciate you doing this. Thank you, Senator.

Philip Boyle Thanks

Appendix H: Ransomware preparedness and recovery guides

- CISA & MS-ISAC: https://www.cisa.gov/publication/ransomware-guide
- Coveware: https://www.coveware.com/blog/2019/5/2/ransomware-first-response-guide-what-to-do-in-the-oh-t-moment
- FTC: https://www.ftc.gov/tips-advice/business-center/small-businesses/cybersecurity/ransomware
- HHS HIPAA Cyber-attack Quick Response Infographic: https://www.hhs.gov/sites/default/files/cyber-attack-quick-response-infographic.gif
- Justice   Department: https://www.justice.gov/criminal-ccips/file/872771/download
- SANS: https://www.sans.org/reading-room/whitepapers/threats/paper/39965
- US   Secret   Service: https://www.secretservice.gov/investigation/Preparing-for-a-Cyber-Incident
- ENISA: https://www.enisa.europa.eu/publications/ransomware
- IBM: https://www.ibm.com/downloads/cas/EV6NAQR4
- EMSISOFT: https://blog.emsisoft.com/en/36921/8-critical-steps-to-take-after-a-ransomware-attack-ransomware-response-guide-for-businesses
- IC3: https://www.ic3.gov/Media/Y2019/PSA191002
- CIS: https://www.cisecurity.org/white-papers/security-primer-ransomware/

Appendix I: Ransomware List by ID Ransomware/Malware Hunter Team (3/3/2021)

Sharing to show that there is more ransomware than just the ones in the headlines. Source: https://id-ransomware.malwarehunterteam.com/

This service currently detects **979** different ransomwares. Here is a complete, dynamic list of what is currently detected:

$$$ Ransomware, 010001, 0kilobypt, 16x, 24H Ransomware, 32aa, 4rw5w, 5ss5c, 777, 7ev3n, 7h9r, 7zipper, 8lock8, AAC, ABCLocker, ACCDFISA v2.0, Adam-Locker, Adhubllka, AES_KEY_GEN_ASSIST, AES-Matrix, AES-NI, AES256-06, AESMew, Afrodita, AgeLocker, Ako / MedusaReborn, Al-Nam-

rood, Al-Namrood 2.0, Alcatraz, Alfa, Allcry, Alma Locker, Alpha, AMBA, Amnesia, Amnesia2, Anatova, AnDROid, AngryDuck, Annabelle 2.1, AnteFrigus, Anubi, Anubis, AnubisCrypt, Apocalypse, Apocalypse (New Variant), ApocalypseVM, ApolloLocker, AresCrypt, Argus, Aris Locker, Armage, ArmaLocky, Arsium, ASN1 Encoder, Ataware, Atchbo, **Aurora**, AutoLocky, AutoWannaCryV2, Avaddon, AVCrypt, Avest, AWT, AxCrypter, aZaZeL, B2DR, Babaxed, **Babuk**, BadBlock, BadEncript, BadRabbit, Bam!, BananaCrypt, BandarChor, Banks1, BarakaTeam, Bart, Bart v2.0, Basilisque Locker, BB Ransomware, BeijingCrypt, BetaSup, BigBobRoss, **BigLock**, Bisquilla, BitCrypt, BitCrypt 2.0, BitCryptor, BitKangoroo, Bitpaymer / DoppelPaymer, BitPyLock, Bitshifter, BitStak, BKRansomware, Black Claw, Black Feather, Black Shades, BlackHeart, BlackKingdom, Blackout, BlackRuby, Blind, Blind 2, Blocatto, BlockFile12, Blooper, Blue Blackmail, Bonsoir, BoooamCrypt, Booyah, BrainCrypt, Brazilian Ransomware, Brick, BrickR, BTCamant, BTCWare, BTCWare Aleta, BTCWare Gryphon, BTCWare Master, BTCWare PayDay, Bubble, Bucbi, Bud, Bug, BugWare, BuyUnlockCode, c0hen Locker, Cancer, Cassetto, Cerber, Cerber 2.0, Cerber 3.0, Cerber 4.0 / 5.0, CerberTear, CheckMail7, Chekyshka, ChernoLocker, Chimera, ChinaJm, ChinaYunLong, ChineseRarypt, CHIP, ClicoCrypter, Clop, Clouded, CmdRansomware, CNHelp, CobraLocker, CockBlocker, Coin Locker, CoinVault, Comrade Circle, Conficker, Consciousness, Conti, CoronaVirus, CorruptCrypt, Cossy, Coverton, Cr1ptT0r Ransomware, CradleCore, CreamPie, Creeper, Cripton, Cripton7zp, Cry128, Cry36, Cry9, Cryakl, CryCryptor, CryFile, CryLocker, CrypMic, Crypren, Crypt0, Crypt0L0cker, Crypt0r, Crypt12, Crypt32, Crypt38, Crypt3r, CryptConsole, CryptConsole3, CryptGh0st, CryptInfinite, CryptoDarkRubix, CryptoDefense, CryptoDevil, CryptoFinancial, CryptoFortress, CryptoGod, CryptoHasYou, CryptoHitman, CryptoJacky, CryptoJoker, CryptoLocker3, CryptoLockerEU, CryptoLocky, CryptoLuck, CryptoMix, CryptoMix Revenge, CryptoMix Wallet, CryptON, Crypton, CryptoPatronum, CryptoPokemon, CryptorBit, CryptoRoger, CryptoShield, CryptoShocker, CryptoTorLocker, CryptoViki, CryptoWall 2.0, CryptoWall 3.0, **CryptoWall 4.0**, CryptoWire, CryptXXX, CryptXXX 2.0, CryptXXX 3.0, CryptXXX 4.0, CryPy, CrySiS, Crystal, CSP Ransomware, CTB-Faker, CTB-Locker, Cuba, CXK-NMSL, Cyborg, Cyrat, D00mEd, Dablio, Damage, DarkoderCryptor, **DarkSide**, DataKeeper, DavesSmith / Balaclava, Dcrtr, DCry, DCry 2.0, Deadly, DeathHiddenTear, DeathHiddenTear v2, DeathNote, DeathOfShadow, DeathRansom, DEcovid19, Decr1pt, DecryptIomega, DecYourData, DEDCryptor, Defender, Defray, Defray777 / RansomEXX, DeriaLock, DeroHE, Desync, **Dharma (.cezar Family)**, Dharma (.dharma Family), Dharma (.onion Family), Dharma (.wallet Family), Digisom, DilmaLocker, DirtyDecrypt, Dishwasher, District, DMA Locker, DMA Locker 3.0, DMA Locker 4.0, DMALocker Imposter, DoggeWiper, Domino, Done, DoNotChange, Donut, DoubleLocker, DriedSister, DryCry, DualShot, Dviide, DVPN, DXXD, DynACrypt, eBayWall, eCh0raix / QNAPCrypt, ECLR Ransomware, EdgeLocker, EduCrypt, EggLocker, Egregor, El Polocker, Enc1, Encrp, EnCrypt, EncryptedBatch, EncrypTile, EncryptoJJS, Encryptor RaaS, Enigma, Enjey Crypter,

EnkripsiPC, EOEO, Epsilon, Erebus, Erica Ransomware, Eris, Estemani, Eternal, Everbe, Everbe 2.0, Everbe 3.0, Evil, Executioner, ExecutionerPlus, Exerwa CTF, Exocrypt XTC, Exorcist Ransomware, Exotic, Extortion Scam, Extractor, EyeCry, Fabiansomware, Fadesoft, Fantom, FartPlz, FCPRansomware, FCrypt, FCT, FenixLocker, FenixLocker 2.0, Fenrir, FilesLocker, FindZip, FireCrypt, Flamingo, Flatcher3, FLKR, FlowEncrypt, Flyper, FonixCrypter, FreeMe, FrozrLock, FRSRansomware, FS0ciety, FTCode, FunFact, FuxSocy Encryptor, Galacti-Crypter, GandCrab, GandCrab v4.0 / v5.0, GandCrab2, GarrantyDecrypt, GC47, Geneve, Gerber, GermanWiper, GetCrypt, GhostCrypt, GhostHammer, Gibberish, Gibon, Gladius, Globe, Globe (Broken), Globe3, GlobeImposter, GlobeImposter 2.0, GoCryptoLocker, Godra, GOG, GoGoogle, GoGoogle 2.0, Golden Axe, GoldenEye, Gomasom, Good, GoRansom, Gorgon, Gotcha, GPAA, GPCode, GPGQwerty, GusCrypter, GX40, Hades, HadesLocker, **Hakbit**, Halloware, Hansom, HappyDayzz, hc6, hc7, HDDCryptor, HDMR, HE-HELP, Heimdall, Hello (WickrMe), HelloKitty, HellsRansomware, Help50, HelpDCFile, Herbst, Hermes, Hermes 2.0, Hermes 2.1, Hermes837, Heropoint, Hi Buddy!, HiddenTear, HildaCrypt, HKCrypt, HollyCrypt, HolyCrypt, HowAreYou, HPE iLO Ransomware, HR, Hucky, Hydra, HydraCrypt, IEncrypt, IFN643, ILElection2020, Ims00ry, ImSorry, Incanto, InducVirus, InfiniteTear, InfinityLock, InfoDot, InsaneCrypt, **IQ**, iRansom, Iron, Ironcat, Ishtar, Israbye, iTunesDecrypt, JabaCrypter, Jack.Pot, Jaff, Jager, JapanLocker, JavaLocker, JCrypt, JeepersCrypt, Jemd, Jigsaw, JNEC.a, JobCrypter, JoeGo Ransomware, JosepCrypt, JSWorm, JSWorm 2.0, JSWorm 4.0, JuicyLemon, JungleSec, Kaenlupuf, Kali, Karma, Karmen, Karo, Kasiski, Katyusha, KawaiiLocker, KCW, Kee Ransomware, KeRanger, Kerkoporta, KesLan, KeyBTC, KEYHolder, KillerLocker, KillRabbit, KimcilWare, Kirk, Knot, KokoKrypt, Kolobo, Kostya, Kozy.Jozy, Kraken, Kraken Cryptor, KratosCrypt, Krider, Kriptovor, KryptoLocker, Kupidon, L33TAF Locker, Ladon, Lalabitch, LambdaLocker, LeakThemAll, LeChiffre, LightningCrypt, Lilocked, Lime, Litra, LittleFinger, LLTP, LMAOxUS, Lock2017, Lock2Bits, Lock93, LockBit, LockBox, LockCrypt, LockCrypt 2.0, LockDown, Locked-In, LockedByte, LockeR, LockerGoga, LockLock, LockMe, Lockout, LockTaiwan, Locky, Loki, Lola, LolKek, LongTermMemoryLoss, LonleyCrypt, LooCipher, Lortok, Lost_Files, LoveServer, LowLevel04, LuciferCrypt, Lucky, MadBit, MAFIA, MafiaWare, Magic, **Magniber**, Major, Makop, Maktub Locker, MalwareTech's CTF, MaMoCrypter, Maoloa, Mapo, Marduk, Marlboro, MarraCrypt, Mars, MarsJoke, Matrix, MauriGo, MaxiCrypt, Maykolin, Maysomware, Maze Ransomware, MCrypt2018, MedusaLocker, MegaCortex, MegaLocker, Mespinoza, Meteoritan, Mew767, Mikoyan, MindSystem, Minotaur, MirCop, MireWare, Mischa, MMM, MNS CryptoLocker, Mobef, MongoLock, Montserrat, MoonCrypter, MorrisBatchCrypt, MOTD, MountLocker, MoWare, MRCR1, MrDec, Muhstik, Mystic, n1n1n1, NanoLocker, NAS Data Compromiser, NCrypt, Nefilim, NegozI, Nemty, Nemty 2.x, Nemty Special Edition, Nemucod, Nemucod-7z, Nemucod-AES, NETCrypton, Netix, Netwalker (Mailto), NewHT, NextCry, Nhtnwcuf, NM4, NMoreira, NMoreira 2.0, Noblis, Nomikon, Non-

Ransomware, NotAHero, Nozelesn, NSB Ransomware, Nuke, NullByte, NxRansomware, Nyton, ODCODC, OhNo!, OmniSphere, OnyxLocker, OoPS, OopsLocker, OpenToYou, OpJerusalem, Ordinypt, Osno, Ouroboros v6, Out-Crypt, OzozaLocker, PadCrypt, Panther, Paradise, Paradise .NET, Paradise B29, **Parasite**, Pay2Key, Paymen45, PayPalGenerator2019, PaySafeGen, PClock, PClock (Updated), PEC 2017, Pendor, Petna, PewCrypt, PewPew, PGPSnippet, PhantomChina, Philadelphia, **Phobos**, PhoneNumber, Pickles, PL Ransomware, Plague17, Planetary Ransomware, PoisonFang, Pojie, PonyFinal, PopCornTime, Potato, Povlsomware, PowerLocky, PowerShell Locker, PowerWare, PPDDDP, Pr0tector, Predator, PrincessLocker, PrincessLocker 2.0, PrincessLocker Evolu-tion, Project23, Project34, Project57, ProLock, Protected Ransomware, Psh-Crypt, PUBG Ransomware, PureLocker, PwndLocker, PyCL, PyCL, PyL33T, PyLocky, qkG, QP Ransomware, QuakeWay, Quimera Crypter, QwertyCrypt, Qweuirtksd, R980, RAA-SEP, RabbitFox, RabbitWare, RackCrypt, Radamant, Radamant v2.1, Radiation, RagnarLocker, RagnarLocker 2.0+, Ragnarok, Ran-dom6, RandomLocker, RandomRansom, Ranion, RanRan, RanRans, Rans0mLocked, RansomCuck, Ransomnix, RansomPlus, Ransomwared, Ran-somWarrior, **Rapid**, Rapid 2.0 / 3.0, RaRansomware, RarVault, Razy, RedBoot, RedEye, RedRum / Tycoon 1.0, RegretLocker, REKTLocker, Rektware, Relock, RemindMe, RenLocker, RensenWare, RetMyData, REvil / Sodinokibi, Reypt-son, Rhino, RNS, RobbinHood, Roga, Rokku, Rontok, RoshaLock, RotorCrypt, Roza, RSA-NI, RSA2048Pro, RSAUtil, Ruby, Russenger, Russian EDA2, **Ryuk**, SAD, SadComputer, Sadogo, SADStory, Sage 2.0, Salsa, SamSam, Sanction, Sanctions, SantaCrypt, Satan, **Satana**, SatanCryptor, Saturn, SaveTheQueen, Scarab, ScareCrow, SD 1.1, Sekhmet, Seon, Sepsis, SerbRansom, Serpent, SFile, ShellLocker, Shifr, Shigo, ShinigamiLocker, ShinoLocker, ShivaGood, Shkolo-taCrypt, Shrug, Shrug2, Shujin, Shutdown57, SifreCozucu, Sifreli, Sigma, Sig-run, SilentDeath, SilentSpring, Silvertor, Simple_Encoder, SintaLocker, Skull Ransomware, SkyFile, SkyStars, Smaug, Smrss32, Snake (Ekans), SnakeLocker, SnapDragon, Snatch, SNSLocker, Solider, Solo Ransomware, Solve, Somik1, Spartacus, SpartCrypt, Spectre, Spider, Spora, Sport, SQ_, Stampado, Stinger, **STOP (Djvu)**, STOP / KEYPASS, StorageCrypter, Storm, Striked, Stroman, Stupid Ransomware, Styx, Such_Crypt, SunCrypt, SuperB, SuperCrypt, Sur-prise, SynAck, SyncCrypt, Syrk, SYSDOWN, SystemCrypter, SZFLocker, Szymekk, T1Happy, TapPiF, Team XRat, Telecrypt, TellYouThePass, Termite, TeslaCrypt 0.x, TeslaCrypt 2.x, TeslaCrypt 3.0, TeslaCrypt 4.0, Teslarvng, Tesla-Ware, TFlower, Thanatos, Thanos, The DMR, TheDarkEncryptor, THIEFQuest, THT Ransomware, ThunderCrypt, ThunderX, tk, Tongda, Torchwood, Total-WipeOut, TowerWeb, ToxCrypt, Tripoli, Trojan.Encoder.6491, Troldesh / Shade, Tron, TrueCrypter, TrumpLocker, TurkStatik, Tycoon 2.0 / 3.0, UCCU, UIWIX, Ukash, UmbreCrypt, UnblockUPC, Ungluk, Unit09, Unknown Crypted, Unknown Lock, Unknown XTBL, Unlock26, Unlock92, Unlock92 2.0, Unlock92 Zipper, UnluckyWare, Useless Disk, UselessFiles, UserFilesLocker, USR0, Uyari, V8Locker, Vaggen, Vapor v1, Vash-Sorena, VaultCrypt, vCrypt, VCrypt, Vega / Jamper / Buran, Velso, Vendetta, VenisRansomware, Venus-

Locker, VHD Ransomware, ViACrypt, VindowsLocker, VisionCrypt, VMola, **VoidCrypt**, Vortex, Vovalex, Vurten, VxLock, Waffle, Waiting, Waldo, Wanna-Cash, WannaCash 2.0, WannaCry, WannaCry.NET, WannaCryFake, WannaCry-OnClick, WannaDie, WannaPeace, WannaRen, WannaScream, WannaSmile, WannaSpam, WastedBit, WastedLocker, Wesker, WhiteRose, WildFire Locker, WininiCrypt, Winnix Cryptor, WinRarer, WonderCrypter, WoodRat, Wooly, Wulfric, X Locker 5.0, XCry, XCrypt, XData, XerXes, XiaoBa, XiaoBa 2.0, XMRLocker, Xorist, Xort, XRTN, XTP Locker 5.0, XYZWare, Yatron, Yogy-nicof, YourRansom, Yyto, Z3, ZariqaCrypt, zCrypt, Zekwacrypt, Zenis, Zeoti-cus, Zeoticus 2.0, **Zeppelin**, ZeroCrypt, Zeronine, Zeropadypt, Zeropadypt NextGen / Ouroboros, ZeroRansom, Zhen, Ziggy, Zilla, ZimbraCryptor, ZinoCrypt, ZipLocker, Zipper, Zoldon, Zorab, ZQ, Zyklon

Appendix J: Survey

Default Report
2020 Ransomware Survey
January 18th 2021, 9:23 am EST

Q1 – Do You Work for a For-Profit, Non-Profit or Government Entity?

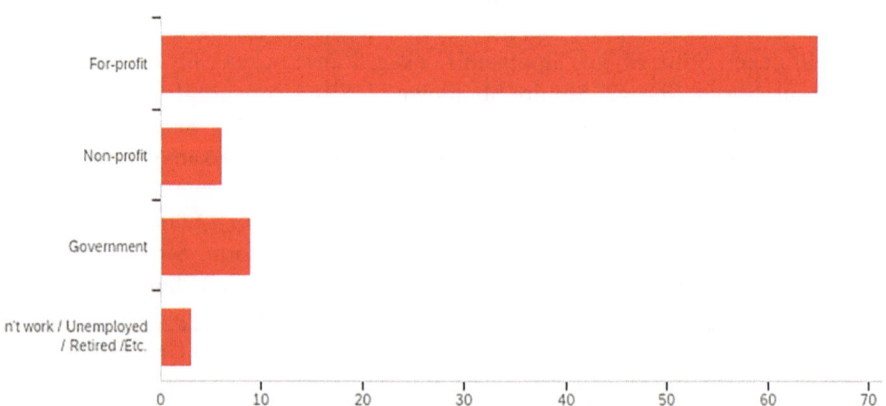

#	Field	Minimum	Maximum	Mean	Std Deviation	Variance	Count
1	Do you work for a for-profit, non-profit or government entity?	1.00	4.00	1.40	0.82	0.67	83

#	Answer	%	Count
1	For-profit	78.31%	65
2	Non-profit	7.23%	6
3	Government	10.84%	9
4	Don't work / Unemployed / Retired /Etc.	3.61%	3
	Total	100%	83

Q30 – In Your Role at Your Organization, Are You A

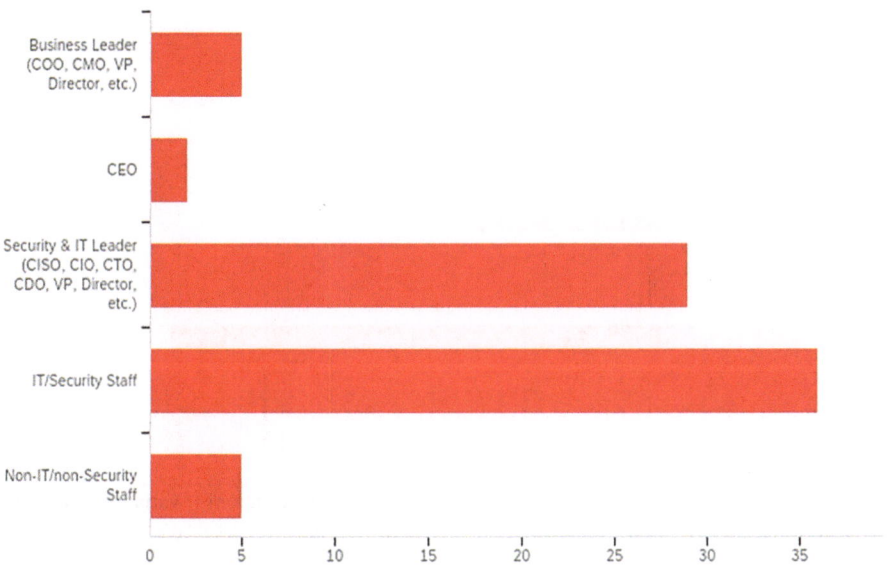

#	Field	Minimum	Maximum	Mean	Std Deviation	Variance	Count
1	In your role at your organization, are you a:	1.00	8.00	6.22	1.55	2.41	77

#	Answer	%	Count
1	Business Leader (COO, CMO, VP, Director, etc.)	6.49%	5
4	CEO	2.60%	2
6	Security & IT Leader (CISO, CIO, CTO, CDO, VP, Director, etc.)	37.66%	29
7	IT/Security Staff	46.75%	36
8	Non-IT/non-Security Staff	6.49%	5
	Total	100%	77

Q2 – How Many Users Are at Your Organization?

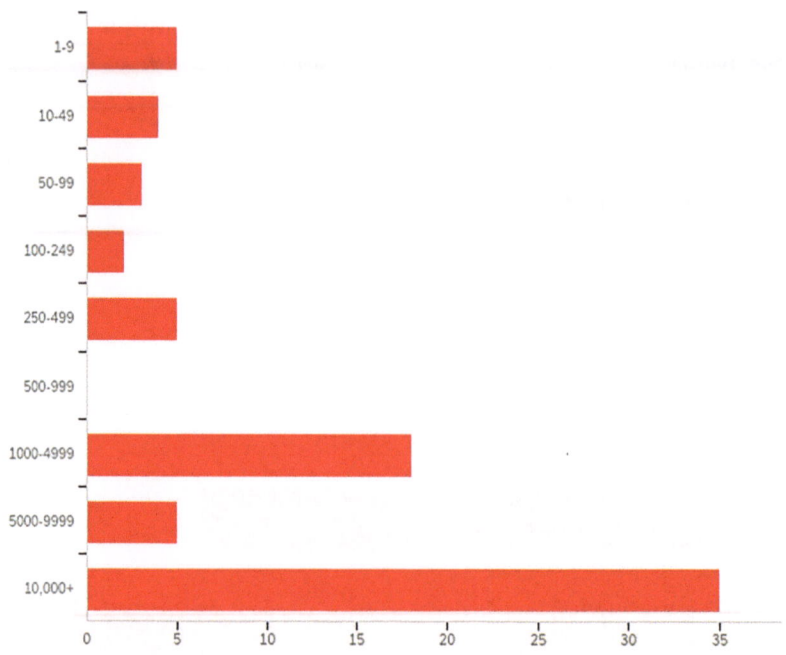

#	Field	Minimum	Maximum	Mean	Std Deviation	Variance	Count
1	How many users are at your organization?	4.00	12.00	9.96	2.58	6.63	77

#	Answer	%	Count
4	1–9	6.49%	5
5	10–49	5.19%	4
6	50–99	3.90%	3
7	100–249	2.60%	2
8	250–499	6.49%	5
9	500–999	0.00%	0
10	1000–4999	23.38%	18
11	5000–9999	6.49%	5
12	10,000+	45.45%	35
	Total	100%	77

Q3 – What Industry Is Your Organization In?

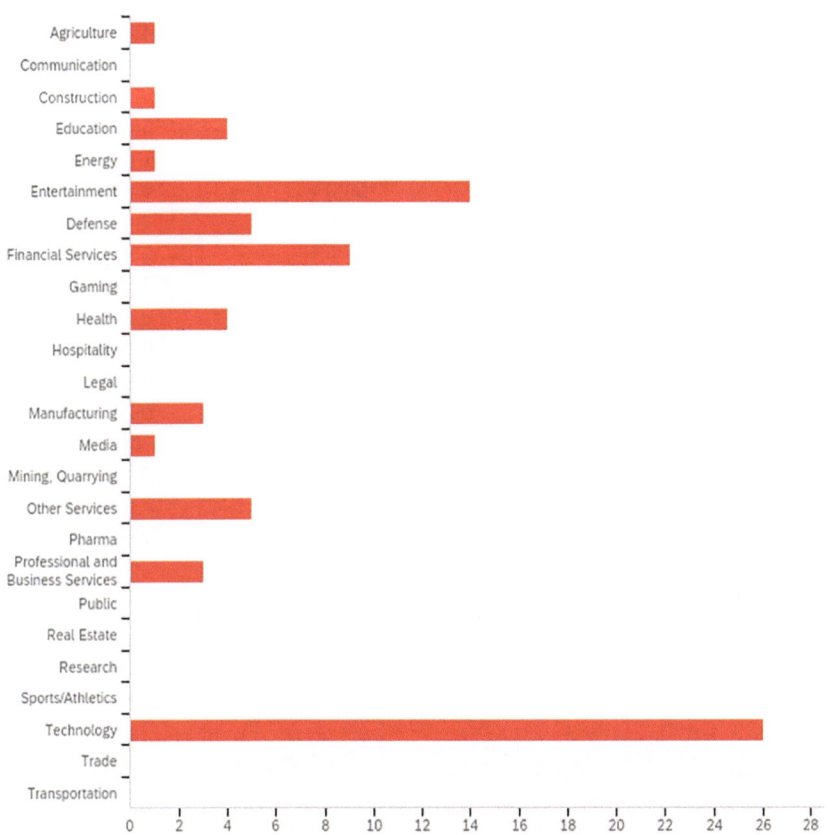

#	Field	Minimum	Maximum	Mean	Std Deviation	Variance	Count
1	What industry is your organization in?	1.00	23.00	13.52	7.61	57.91	77

#	Answer	%	Count
1	Agriculture	1.30%	1
2	Communication	0.00%	0
3	Construction	1.30%	1
4	Education	5.19%	4
5	Energy	1.30%	1
6	Entertainment	18.18%	14
7	Defense	6.49%	5

8	Financial Services	11.69%	9
9	Gaming	0.00%	0
10	Health	5.19%	4
11	Hospitality	0.00%	0
12	Legal	0.00%	0
13	Manufacturing	3.90%	3
14	Media	1.30%	1
15	Mining, Quarrying	0.00%	0
16	Other Services	6.49%	5
17	Pharma	0.00%	0
18	Professional and Business Services	3.90%	3
19	Public	0.00%	0
20	Real Estate	0.00%	0
21	Research	0.00%	0
22	Sports/Athletics	0.00%	0
23	Technology	33.77%	26
24	Trade	0.00%	0
25	Transportation	0.00%	0
	Total	100%	77

Q5 – Does Your Organization Have an Incident Response Plan?

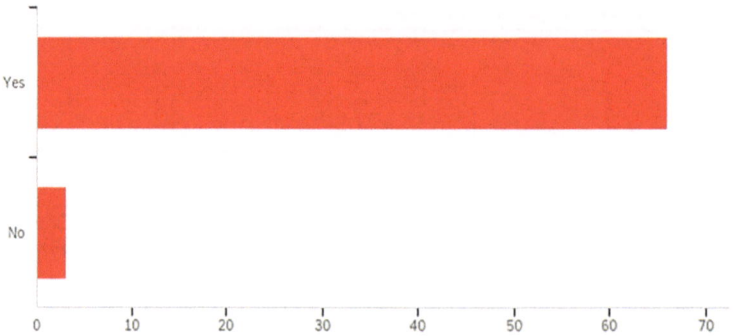

#	Field	Minimum	Maximum	Mean	Std Deviation	Variance	Count
1	Does your organization have an incident response plan?	1.00	2.00	1.04	0.20	0.04	69

#	Answer	%	Count
1	Yes	95.65%	66
2	No	4.35%	3
	Total	100%	69

Q6 – Does Your Organization Have an Incident Response Plan for Ransomware?

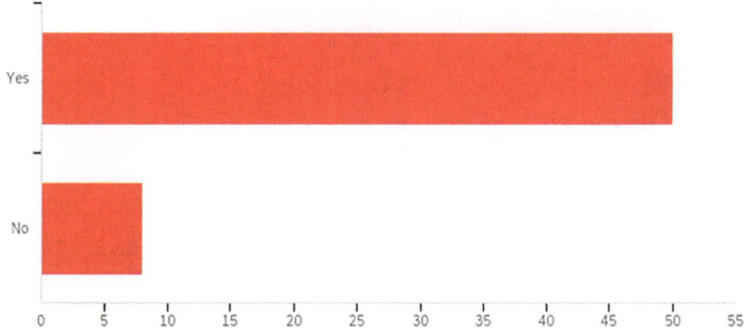

#	Field	Minimum	Maximum	Mean	Std Deviation	Variance	Count
1	Does your organization have an incident response plan for ransomware?	1.00	2.00	1.14	0.34	0.12	58

#	Answer	%	Count
1	Yes	86.21%	50
2	No	13.79%	8
	Total	100%	58

Q7 – Does Your Organization Have Bitcoin Set Aside for Ransomware Payments? (Not via Insurance)

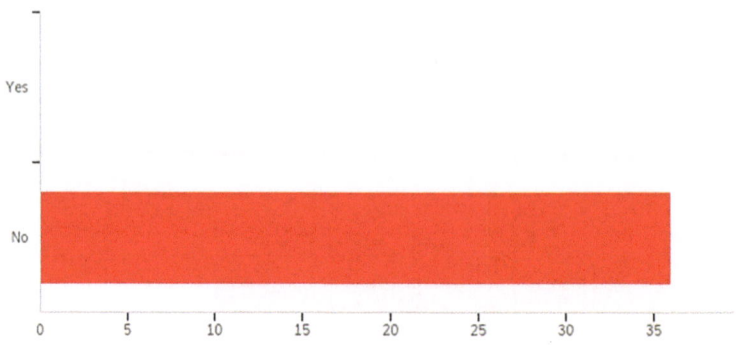

#	Field	Minimum	Maximum	Mean	Std Deviation	Variance	Count
1	Does your organization have bitcoin set aside for ransomware payments? (Not via insurance)	2.00	2.00	2.00	0.00	0.00	36

#	Answer	%	Count
1	Yes	0.00%	0
2	No	100.00%	36
	Total	100%	36

Q8 – Does Your Organization Have a Contract with a Third Party Negotiator in Case of the Need to Negotiate Ransoms?

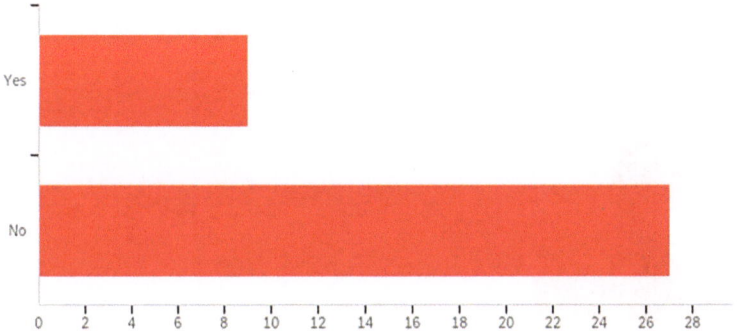

#	Field	Minimum	Maximum	Mean	Std Deviation	Variance	Count
1	Does your organization have a contract with a third party negotiator in case of the need to negotiate ransoms?	1.00	2.00	1.75	0.43	0.19	36

#	Answer	%	Count
1	Yes	25.00%	9
2	No	75.00%	27
	Total	100%	36

Q11 – Would Your Organization Pay a Ransom for any of the Following Reasons

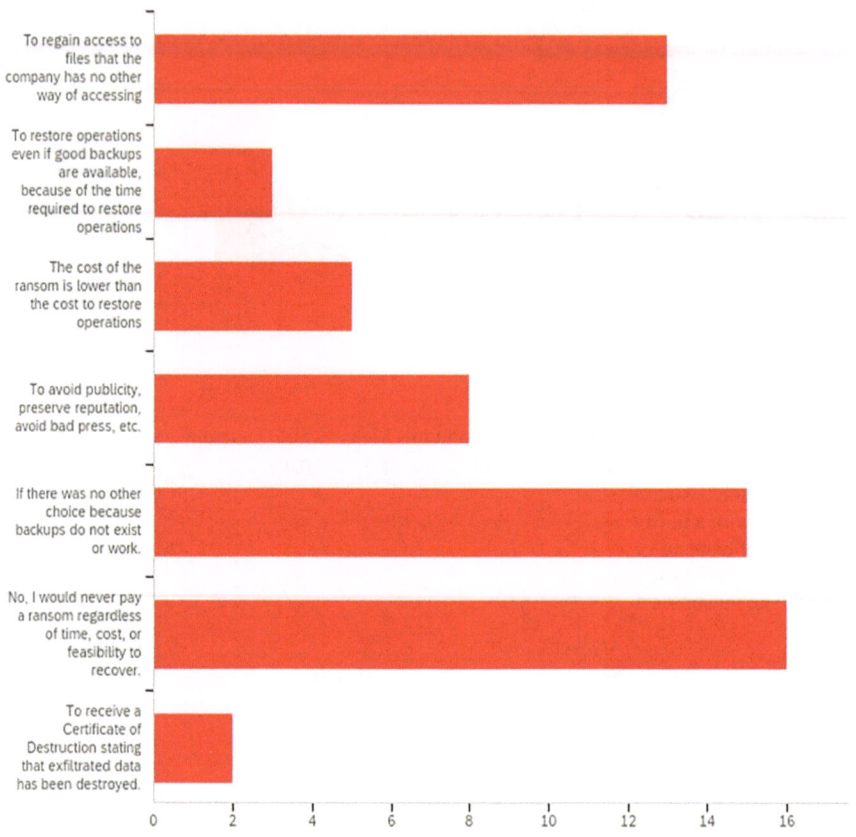

#	Answer	%	Count
1	To regain access to files that the company has no other way of accessing	20.97%	13
2	To restore operations even if good backups are available, because of the time required to restore operations	4.84%	3
3	The cost of the ransom is lower than the cost to restore operations	8.06%	5
4	To avoid publicity, preserve reputation, avoid bad press, etc.	12.90%	8

5	If there was no other choice because backups do not exist or work.	24.19%	15
6	No, I would never pay a ransom regardless of time, cost, or feasibility to recover.	25.81%	16
8	To receive a Certificate of Destruction stating that exfiltrated data has been destroyed.	3.23%	2
	Total	100%	62

Q12 – Would Your Organization Pay a "Reasonable" Ransom Demand if it Would Take Longer Than a Certain Amount of Time to Recover Operations?

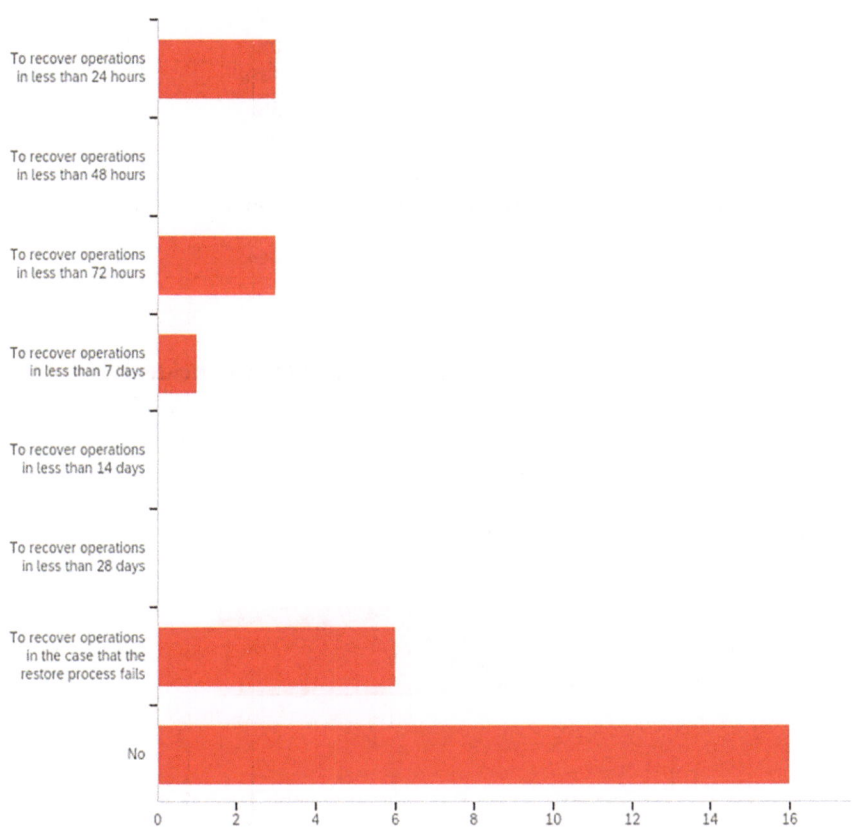

#	Field	Minimum	Maximum	Mean	Std Deviation	Variance	Count
1	Would your organization pay a "reasonable" ransom demand if it would take longer than a certain amount of time to recover operations?	1.00	8.00	6.41	2.43	5.90	29

#	Answer	%	Count
1	To recover operations in less than 24 h	10.34%	3
2	To recover operations in less than 48 h	0.00%	0
3	To recover operations in less than 72 h	10.34%	3
4	To recover operations in less than 7 days	3.45%	1
5	To recover operations in less than 14 days	0.00%	0
6	To recover operations in less than 28 days	0.00%	0
7	To recover operations in the case that the restore process fails	20.69%	6
8	No	55.17%	16
	Total	100%	29

Q13 – Has Your Organization Experienced a Ransomware Incident?

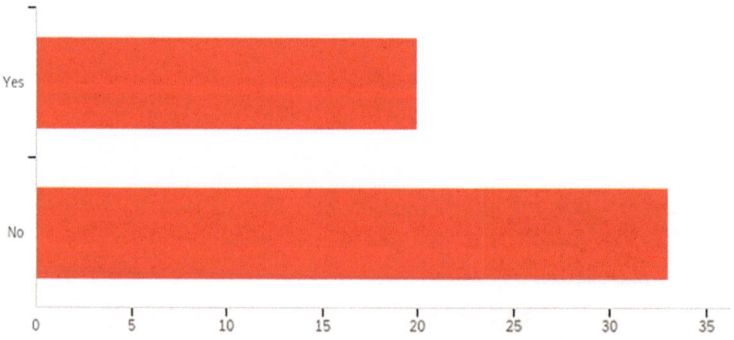

#	Field	Minimum	Maximum	Mean	Std Deviation	Variance	Count
1	Has your organization experienced a ransomware incident?	1.00	2.00	1.62	0.48	0.23	53

#	Answer	%	Count
1	Yes	37.74%	20
2	No	62.26%	33
	Total	100%	53

Q31 – Did You Discuss the Incident with Anyone Outside the Organization That Wasn't Directly Responsible for Investigating or Remediating the Incident? If So, Who?

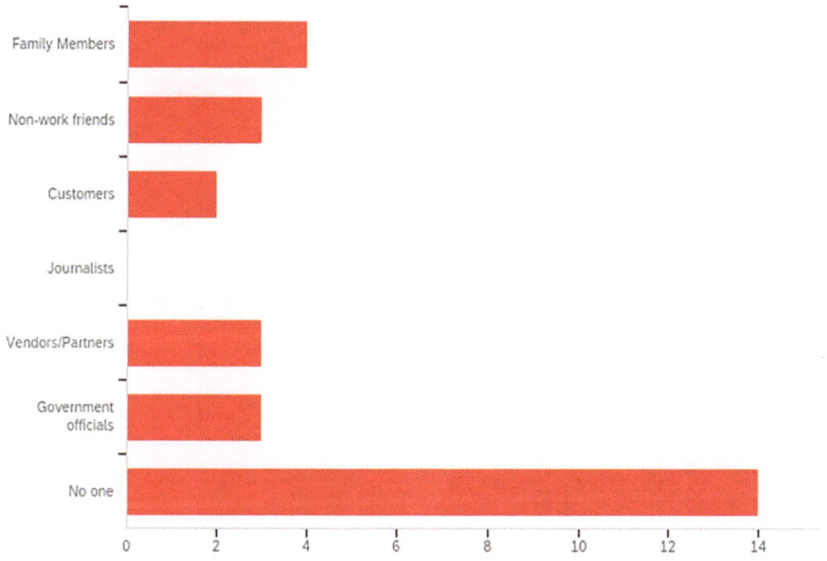

#	Answer	%	Count
1	Family Members	13.79%	4
2	Non-work friends	10.34%	3
3	Customers	6.90%	2
4	Journalists	0.00%	0
5	Vendors/Partners	10.34%	3
6	Government officials	10.34%	3
8	No one	48.28%	14
	Total	100%	29

Q14 – Did Your Organization Pay a Ransom from Its Own Funds, a Third Party Service (ex. Insurance), or Both?

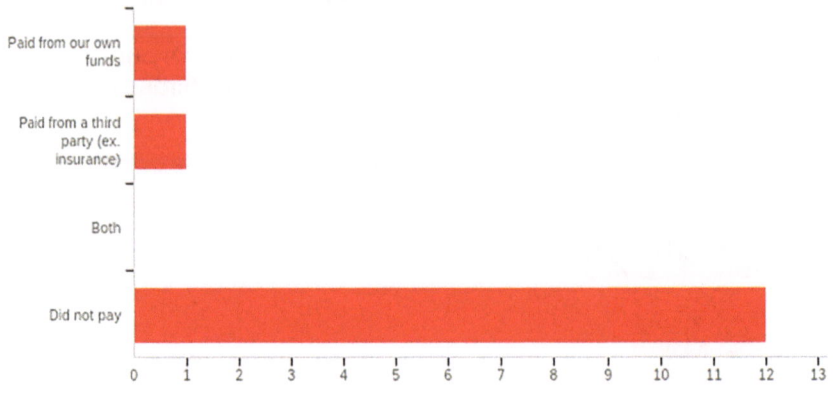

#	Field	Minimum	Maximum	Mean	Std Deviation	Variance	Count
1	Did your organization pay a ransom from its own funds, a third party service (ex. Insurance), or both?	1.00	6.00	5.36	1.59	2.52	14

#	Answer	%	Count
1	Paid from our own funds	7.14%	1
2	Paid from a third party (ex. insurance)	7.14%	1
3	Both	0.00%	0
6	Did not pay	85.71%	12
	Total	100%	14

Q16 – Was the Final Ransom Payment More, Equal, or Less Than What Was Originally Requested by the Ransomware Operator?

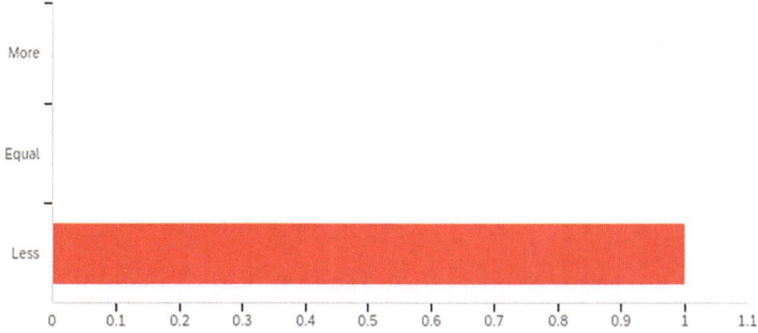

#	Field	Minimum	Maximum	Mean	Std Deviation	Variance	Count
1	Was the final ransom payment more, equal, or less than what was originally requested by the ransomware operator?	3.00	3.00	3.00	0.00	0.00	1

#	Answer	%	Count
1	More	0.00%	0
2	Equal	0.00%	0
3	Less	100.00%	1
	Total	100%	1

Q23 – Did the Decryption Process Work?

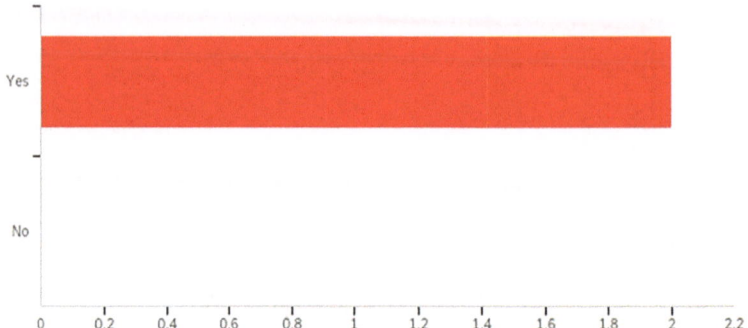

#	Field	Minimum	Maximum	Mean	Std Deviation	Variance	Count
1	Did the decryption process work?	1.00	1.00	1.00	0.00	0.00	2

#	Answer	%	Count
1	Yes	100.00%	2
2	No	0.00%	0
	Total	100%	2

Q24 – What Percentage of Files Were Decrypted? (Estimate. Skip if You Don't Know)

#	Field	Minimum	Maximum	Mean	Std Deviation	Variance	Count
1	1	90.00	100.00	95.00	5.00	25.00	2

Q17 – Was There Unrecoverable Data After the Decryption Process was Applied?

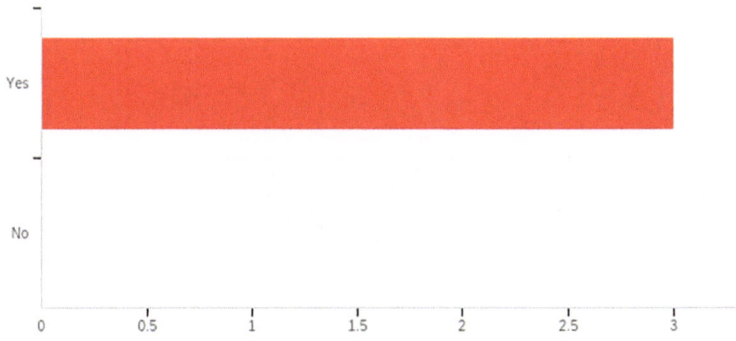

#	Field	Minimum	Maximum	Mean	Std Deviation	Variance	Count
1	Was there unrecoverable data after the decryption process was applied?	1.00	1.00	1.00	0.00	0.00	3

#	Answer	%	Count
1	Yes	100.00%	3
2	No	0.00%	0
	Total	100%	3

Q18 – Did You Have to Engage with the Ransomware "Customer Service?"

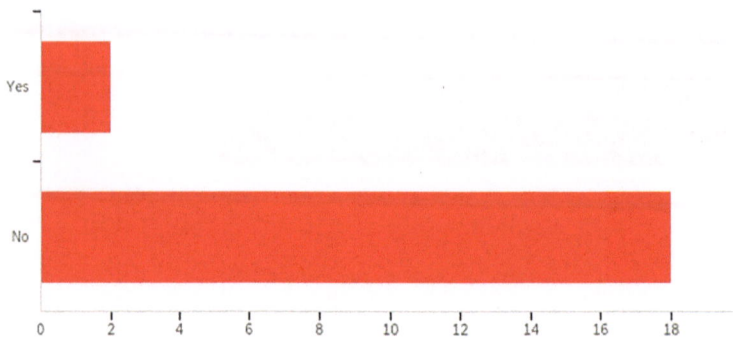

#	Field	Minimum	Maximum	Mean	Std Deviation	Variance	Count
1	Did you have to engage with the Ransomware "customer service?"	1.00	2.00	1.90	0.30	0.09	20

#	Answer	%	Count
1	Yes	10.00%	2
2	No	90.00%	18
	Total	100%	20

Q19 – How Did You Engage with Ransomware Customer Service?

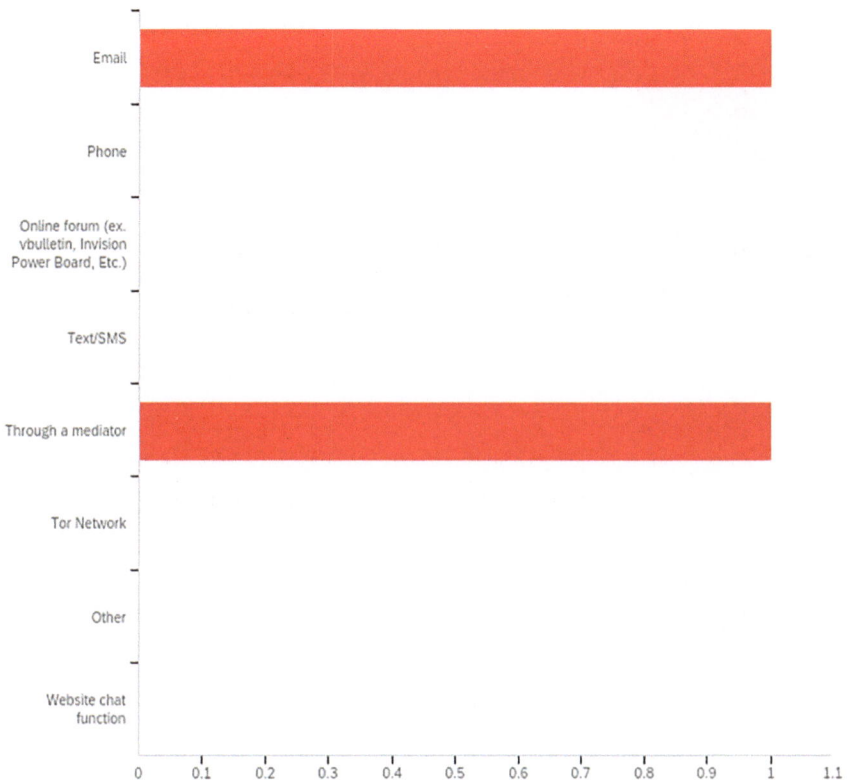

#	Answer	%	Count
1	Email	50.00%	1
2	Phone	0.00%	0
3	Online forum (ex. vbulletin, Invision Power Board, Etc.)	0.00%	0
4	Text/SMS	0.00%	0
5	Through a mediator	50.00%	1
7	Tor Network	0.00%	0
8	Other	0.00%	0
9	Website chat function	0.00%	0
	Total	100%	2

Q20 – Was There Another Ransomware Incident at the Same Organization Within 12 Months of Paying the Ransom?

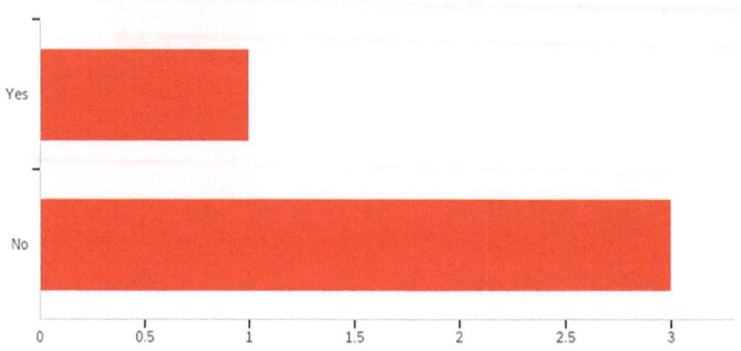

#	Field	Minimum	Maximum	Mean	Std Deviation	Variance	Count
1	Was there another ransomware incident at the same organization within 12 months of paying the ransom?	1.00	2.00	1.75	0.43	0.19	4

#	Answer	%	Count
1	Yes	25.00%	1
2	No	75.00%	3
	Total	100%	4

Bibliography

Golabek-Goldman, Michele. 2014. "A NEW STRATEGY FOR REDUCING THE THREAT OF DANGEROUS ØDAY SALES TO GLOBAL SECURITY AND THE ECONOMY." *SSRN*. March 25. Accessed October 23, 2020. https://poseidon01.ssrn.com/delivery. php?ID=829105006026084119088083111113112014051023066041063035095113106104 026106089027028120017118056026029032013092097088093117089004053020057005079 112074064024024068102022093081010007122074117066071008004021096123126.

n.d. "18 U.S. Code § 1030 – Fraud and related activity in connection with computers ." *Cornell Law School Legal Information institute*. Accessed November 20, 2020. https://www.law.cornell.edu/uscode/text/18/1030.

Abrams, Lawrence. 2020a. "Cyber insurer's security scans reduced ransomware claims by 65%." *Bleeping Computer*. September 22. Accessed September 25, 2020. https://www.bleepingcomputer.com/news/security/cyber-insurers-security-scans-reduced-ransomware-claims-by-65-percent/.

———. 2020b. "Egregor ransomware print bombs printers with ransom notes." *Bleeping Computer*. November 18. Accessed 2020 23, November. https://www.bleepingcomputer.com/news/security/egregor-ransomware-print-bombs-printers-with-ransom-notes/.

———. 2020c. "Fake ransomware decryptor double-encrypts desperate victims' files." *Bleeping Computer*. June 6. Accessed December 20, 2020. https://www.bleepingcomputer.com/news/security/fake-ransomware-decryptor-double-encrypts-desperate-victims-files/.

———. 2020d. "Ransomware Attackers Use Your Cloud Backups Against You." *Bleeping Computer*. Marc 3. Accessed December 28, 2020. https://www.bleepingcomputer.com/news/security/ransomware-attackers-use-your-cloud-backups-against-you/.

———. 2020e. "Surge of MongoDB ransom attacks use GDPR as extortion leverage." *Bleeping Computer*. July 2. Accessed December 26, 2020. https://www.bleepingcomputer.com/news/security/surge-of-mongodb-ransom-attacks-use-gdpr-as-extortion-leverage/.

———. 2020f. "Tyler Technologies paid ransomware gang for decryption key." *Bleeping Computer*. October 10. Accessed October 14, 2020. https://www.bleepingcomputer.com/news/security/tyler-technologies-paid-ransomware-gang-for-decryption-key/.

Accenture. 2020. "Shady deals: The destructive relationship between network access sellers and ransomware groups." *Cyber Defense*. October 12. Accessed October 18, 2020. https://www.accenture.com/us-en/blogs/cyber-defense/destructive-relationship-between-network-access-sellers-and-ransomware-groups.

Airgap Networks. 2020. "Zero Trust Security in Action: Airgap Ransomware Kill Switch." *Airgap Networks*. October 23. Accessed January 20, 2021. https://airgapnetworks.medium.com/zero-trust-security-in-action-airgap-ransomware-kill-switch-f46443cee09f.

© The Editor(s) (if applicable) and The Author(s), under exclusive license to
Springer Nature Switzerland AG 2024
H. Halikias, *Digital Shakedown*, https://doi.org/10.1007/978-3-031-65438-1

Akpan, Nsikan. 2019. "Ransomware and data breaches linked to uptick in fatal heart attacks." *PBS News Hour.* October 24. Accessed November 5, 2020. https://www.pbs.org/newshour/science/ransomware-and-other-data-breaches-linked-to-uptick-in-fatal-heart-attacks.

Alder, Steve. 2021. "Rady Children's Hospital Facing Class Action Lawsuit over Blackbaud Ransomware Attack." *HIPAA Journal.* January 26. Accessed February 27, 2021. https://www.hipaajournal.com/rady-childrens-hospital-facing-class-action-lawsuit-over-blackbaud-ransomware-attack/.

American Public Power Association. 2019. "Public Power Cyber Incident Response Playbook." Guidance.   https://www.publicpower.org/system/files/documents/Public-Power-Cyber-Incident-Response-Playbook.pdf.

Anderson, Ross. 2008. *Security Engineering. Second Edition.* Edinburgh: Wiley.

Ang, Carmen. 2021. "Comparing Bitcoin's Market Cap to Other Cryptocurrencies." *Visual Capitalist.* January 13. Accessed January 31, 2021. https://www.visualcapitalist.com/bitcoin-market-cap-compared-to-crypto/.

2017. "ARTICLE 29 DATA PROTECTION WORKING PARTY." Advisory, Brussels, Belgium. http://ec.europa.eu/newsroom/document.cfm?doc_id=47741.

Ashford, Warwick. 2019. "Sodin ransomware exploiting Windows zero-day, Kaspersky warns." *Computer Weekly.* July 3. Accessed November 3, 2020. https://www.computerweekly.com/news/252466220/Sodin-ransomware-exploiting-Windows-zero-day-Kaspersky-warns.

Associated Press. 2005a. "Latest Internet attack holds computer files hostage." *seclists.org.* May 25. Accessed November 14, 2020. https://seclists.org/interesting-people/2005/May/249.

———. 2005b. "Latest Internet attack holds computer files hostage. Breaking News from globeandmail.com." May 24. Accessed November 14, 2020. https://advance-lexis-com.revproxy.brown.edu/api/document?collection=news&id=urn:contentItem:4NGB-XWG0-TXJC-4107-0 0000-00&context=1516831.

Baipal, P., and R. Enbody. 2020. "Memory Forensics Against Ransomware." *2020 International Conference on Cyber Security and Protection of Digital Services (Cyber Security).* Dublin, Ireland: IEEE. 1-8. https://ieeexplore.ieee.org/document/9138853.

BBC News. 2020. "Uber ex-security boss accused of covering up hack attack." *BBC News.* August 21. Accessed January 10, 2021. https://www.bbc.com/news/technology-53861375.

Bernstein, Daniel J. 2008. "The Salsa20 Family of Stream Ciphers." *Springer Link.* Edited by M. Robshaw and O. Billet. Accessed September 24, 2020. https://doi.org/10.1007/978-3-540-68351-3_8.

Beschizza, Rob. 2007. "The New HD-DVD/Blu-Ray Hack: What It Might Mean For Us." *Wired.* February 13. Accessed September 23, 2020. https://www.wired.com/2007/02/the-new-hddvdbl/.

Blum, Raymond. October 22, 2013. "NYC Tech Talk Series: How Google backs up the internet." New York, NY: YouTube. https://www.youtube.com/watch?v=eNliOm9NtCM.

Bohlen, Celestine. 1998. "Italian Ban On Paying Kidnappers Stirs Anger." *The New York Times.* February 1. Accessed December 9, 2020. https://www.nytimes.com/1998/02/01/world/italian-ban-on-paying-kidnappers-stirs-anger.html.

Boyle, Phil, interview by Harry Halikias. 2020a. *NY State Senator* (October 1).

———. 2020b. "Senate Bill S7246 ." *The New York State Senate.* January 14. Accessed February 13, 2020. https://www.nysenate.gov/legislation/bills/2019/s7246.

Bracy, Jedidiah. 2020. "Amid a global pandemic, ransomware increasingly targets hospitals." *IAPP: The Privacy Advisor.* December 1. Accessed January 21, 2021. https://iapp.org/news/a/amid-a-global-pandemic-ransomware-increasingly-targets-hospitals/.

Brandom, Russell. 2017. "Is Microsoft to blame for the largest ransomware attacks in internet history?" *The Verge.* May September. Accessed September 21, 2020. https://www.theverge.com/2017/5/15/15641198/microsoft-ransomware-wannacry-security-patch-upgrade-wannacrypt.

Business Insider. 2020. "India becomes the second-most targeted country for ransomware after surge in attacks over the last three months." *Business Insider India.* October 6. Accessed

February 21, 2021. https://www.businessinsider.in/tech/news/india-becomes-the-most-targeted-country-for-ransomware-after-surge-in-attacks-over-the-last-three-months/article-show/78514670.cms.

1994. *Dumb and Dumber.* Directed by Peter Farrelly. Performed by Jim Carrey.

Chainalysis. 2020. "The Chainalysis 2020 Crypto Crime Report." *Chainalysis.* January. Accessed December 8, 2020. https://go.chainalysis.com/2020-Crypto-Crime-Report.html.

Check Point Research. 2020. "Ransomware Evolved: Double Extortion." *CP<R>.* April 16. Accessed September 18, 2020. https://research.checkpoint.com/2020/ransomware-evolved-double-extortion/.

Chernenko, Elena, Oleg Demidov, and Fyodor Lukyanov. 2018. "Increasing International Cooperation in Cybersecurity and Adapting Cyber Norms." *Council on Foreign Relations.* February 23. Accessed March 25, 2021. https://www.cfr.org/report/increasing-international-cooperation-cybersecurity-and-adapting-cyber-norms.

Cimpanu, Catalin. 2020a. "Cloud provider stopped ransomware attack but had to pay ransom demand anyway." *ZDNet.* July 17. Accessed September 21, 2020. https://www.zdnet.com/article/cloud-provider-stopped-ransomware-attack-but-had-to-pay-ransom-demand-anyway/.

———. 2020b. "Company shuts down because of ransomware, leaves 300 without jobs just before holidays." *ZDNet.* January 3. Accessed January 10, 2021. https://www.zdnet.com/article/company-shuts-down-because-of-ransomware-leaves-300-without-jobs-just-before-holidays/.

———. 2019a. "GandCrab ransomware operation says it's shutting down." *ZDNet.* June 1. Accessed January 23, 2020. https://www.zdnet.com/article/gandcrab-ransomware-operation-says-its-shutting-down/.

———. 2020c. "Hackers tried (and failed) to install ransomware using a zero-day in Sophos firewalls." *ZDNet.* May 21. Accessed November 3, 2020. https://www.zdnet.com/article/hackers-tried-and-failed-to-install-ransomware-using-a-zero-day-in-sophos-firewalls/.

———. 2019b. "Over 500 US schools were hit by ransomware in 2019." *ZDNet.* October 1. Accessed October 5, 2020. https://www.zdnet.com/article/over-500-us-schools-were-hit-by-ransomware-in-2019/.

———. 2019c. "Ransomware incident to cost Danish company a whopping $95 million." *ZDNet.* September 30. Accessed May 21, 2020. https://www.zdnet.com/article/ransomware-incident-to-cost-danish-company-a-whopping-95-million/.

———. 2020d. "Ransomware installs Gigabyte driver to kill antivirus products." *ZDNet.* February 7. Accessed January 21, 2021. https://www.zdnet.com/article/ransomware-installs-gigabyte-driver-to-kill-antivirus-products/.

———. 2020e. "Top exploits used by ransomware gangs are VPN bugs, but RDP still reigns supreme." *ZDNet.* August 24. Accessed October 3, 2020. https://www.zdnet.com/article/top-exploits-used-by-ransomware-gangs-are-vpn-bugs-but-rdp-still-reigns-supreme/.

———. 2020f. "UHS hospital network hit by ransomware attack." *ZDNet.* September 28. Accessed January 21, 2021. https://www.zdnet.com/article/uhs-hospital-network-hit-by-ransomware-attack/.

Ciphertrace. 2021. "Cryptocurrency Crime and Anti-Money Laundering Report, February 2021." *CipherTrace.* February. Accessed December 5, 2021. https://ciphertrace.com/2020-year-end-cryptocurrency-crime-and-anti-money-laundering-report/.

CipherTrace Cryptocurrency Intelligence. 2019. *Cryptocurrency Anti-Money Laundering Report, 2019 Q2.* Ciphertrace. https://ciphertrace.com/wp-content/uploads/2019/09/CipherTrace-Cryptocurrency-Anti-Money-Laundering-Report-2019-Q2-3.pdf.

Cloudflare. n.d. "What Happens in a TLS Handshake? | SSL Handshake." *Cloudflare.* Accessed September 2020, 24. https://www.cloudflare.com/learning/ssl/what-happens-in-a-tls-handshake/.

Cobalt Strike by HelpSystems. n.d. "Beacon Covert C2 Payload." *cobaltstrike.* Accessed August 15, 2020. https://www.cobaltstrike.com/help-beacon.

Code 42. 2018. *2018 Data Exposure Report.* Minneapolis, MN: Code 42. https://hosteddocs.ittool-box.com/Code42-Data-Exposure-Report-2-v11.pdf.

Conboy, Jason. 2020. "Cyber Summit 2020: Day One (Key Cyber Insights)." YouTube. https://www.youtube.com/watch?t=5109&v=yWqN0zLOOVs.

Congressional Research Service. 2020. "Cybercrime and the Law: Computer Fraud and Abuse Act (CFAA) and the 116th Congress." Research, Washington D.C. https://fas.org/sgp/crs/misc/R46536.pdf.

Corder, Mike, and Frank Bajak. 2021. "Cybercops derail malware botnet, FBI makes ransomware arrest." *ABC News.* January 27. Accessed February 13, 2021. https://abcnews.go.com/International/wireStory/cyber-cops-nations-team-disrupt-dangerous-malware-75511983.

Council of Europe. 2001. "Convention on Cybercrime." Budapest, Hungary, November 23. https://rm.coe.int/CoERMPublicCommonSearchServices/DisplayDCTMContent?documentId=0900001680081561.

———. 2020. "The Budapest Convention on Cybercrime: benefits and impact in practice." Analysis, Strasbourg, France. https://rm.coe.int/t-cy-2020-16-bc-benefits-rep-provisional/16809ef6ac.

Coveware. 2020. "Ransomware Demands continue to rise as Data Exfiltration becomes common, and Maze subdues." *Coveware.* November 4. Accessed December 30, 2020. https://www.coveware.com/blog/q3-2020-ransomware-marketplace-report.

Daraghi, Tooska, Ali Dehghantanha, Pooneh Nikkhah Bahrami, Mauro Conti, and Giuseppe Bianchi. 2019. "A Cyber-Kill-Chain based taxonomy of crypto-ransomware features." *Journal of Computer Virology and Hacking Techniques* (Springer) 277–305. https://doi.org/10.1007/s11416-019-00338-7.

Darktrace. 2021. "Threat Landscape Series: Neutralizing Ransomware." *Darktrace.* Accessed January 29, 2023. https://darktrace.com/resources/threat-landscape-series-neutralizing-ransomware.

Datto Inc. 2019. "Cost of Ransomware Related Downtime Increased More Than 200 Percent, an Amount 23 Times Greater Than the Ransom Request; Datto Global Survey Uncovers Business Impact and Increasing Frequency of Ransomware Attacks." *Business Wire.* October 16. Accessed November 5, 2020. https://www.businesswire.com/news/home/20191016005043/en/Cost-of-Ransomware-Related-Downtime-Increased-More-Than-200-Percent-an-Amount-23-Times-Greater-Than-the-Ransom-Request.

Davis, Jessica. 2020. "UPDATE: UHS Health System Confirms All US Sites Affected by Ransomware Attack." *Helth IT Security.* October 3. Accessed November 9, 2020. https://healthitsecurity.com/news/uhs-health-system-confirms-all-us-sites-affected-by-ransomware-attack.

Dignan, Larry. 2019. "Ransomware attacks: Why and when it makes sense to pay the ransom." *ZDNer.* June 27. Accessed January 8, 2020. https://www.zdnet.com/article/why-and-when-it-makes-sense-to-pay-the-ransom-in-ransomware-attacks/.

Dudley, Renee. 2019a. "The Extortion Economy: How Insurance Companies Are Fueling a Rise in Ransomware Attacks." *ProPublica.* August 27. Accessed July 21, 2020. https://www.propublica.org/article/the-extortion-economy-how-insurance-companies-are-fueling-a-rise-in-ransomware-attacks.

———. 2019b. "The New Target That Enables Ransomware Hackers to Paralyze Dozens of Towns and Businesses at Once." *ProPublica.* September 12. Accessed November 5, 2020. https://www.propublica.org/article/the-new-target-that-enables-ransomware-hackers-to-paralyze-dozens-of-towns-and-businesses-at-once.

Dutton, Yvonne M., and Joe Bellish. 2014. "Refusing to Negotiate: Analyzing the Legality and Practicality of a Piracy Ransom Ban." *Cornell International Law Journal* Vol. 4 (Iss. 2, Article 2): 299-328. https://core.ac.uk/download/pdf/216749487.pdf.

EMSISOFT. 2019. "Caution! Ryuk Ransomware decryptor damages larger files, even if you pay." *EMSISOFT Blog.* December 9. Accessed December 23, 2020. https://blog.emsisoft.com/en/35023/bug-in-latest-ryuk-decryptor-may-cause-data-loss/.

———. n.d. "Free Ransomware Decryption Tools." *EMSISOFT.* Accessed January 23, 2021. https://www.emsisoft.com/ransomware-decryption-tools/.

Emsisoft Malware Lab. 2020. "Ransomware surges in education sector in Q3 as attackers wait patiently for start of school year." *Emsisoft Blog.* November 13. Accessed November 14, 2020.

https://blog.emsisoft.com/en/37193/ransomware-surges-in-education-sector-in-q3-as-attackers-wait-patiently-for-start-of-school-year/.

ENISA. October 20, 2020. *ENISA Threat Landscape 2020 – Ransomware*. Analysis, European Union Agency for Cybersecurity.

———. October 20, n.d. "Zero-Day." *Glossary*. Accessed October 25, 2020. https://www.enisa.europa.eu/topics/csirts-in-europe/glossary/zero-day.

Erskine, Toni, and Madeline Carr. 2016. "Beyond 'Quasi-Norms': The Challenges and Potential of Engaging with Norms in Cyberspace." *Osula/Rõigas (Hg.): International cyber norms. Legal, policy & industry perspectives. Tallinn, Estonia: NATO Cooperative Cyber Defence Centre of Excellence* (NATO CCD COE Publications) 87–110.

European Parliament. n.d. "Art. 33 GDPR - Notification of a personal data breach to the supervisory authority." *Intersoft Consulting*. Accessed December 26, 2020. https://gdpr-info.eu/art-33-gdpr/.

———. 2013. "Position of the European Parliament." Consolidated legislative document, Brussels, Belgium. https://www.europarl.europa.eu/doceo/document/TC1-COD-2010-0273_EN.pdf.

EUROPOL. 2020. "No More Ransom: how 4 millions victims of ransomware have fought back against hackers ." *EUROPOL Newsroom*. July 27. Accessed January 11, 2021. https://www.europol.europa.eu/newsroom/news/no-more-ransom-how-4-millions-victims-of-ransomware-have-fought-back-against-hackers.

Falcone, Robert. 2016. "Shamoon 2: Return of the Disttrack Wiper." *Palo Alto Unit 42*. November 30. Accessed September 20, 2020. https://unit42.paloaltonetworks.com/unit42-shamoon-2-return-disttrack-wiper/.

FBI. 2015. "FBI Launches Nationwide Awareness Campaign." *FBI News: Economic Espionage*. July 23. Accessed March 21, 2020. https://www.fbi.gov/news/stories/economic-espionage.

———. n.d. *Ransomware*. Accessed April 9, 2020. https://www.fbi.gov/scams-and-safety/common-scams-and-crimes/ransomware.

Ferillo, Paul A. May 18, 2020. *ISC2 Long Island Chapter Meeting*. Long Island, NY.

FinCEN. 2020. *Advisory on Ransomware and the Use of the Financial System to Facilitate Ransom Payments*. Advisory, Washington D.C.: FinCEN. https://www.fincen.gov/sites/default/files/advisory/2020-10-01/Advisory%20Ransomware%20FINAL%20508.pdf.

FireEye. 2016. "Endpoint Security." Marketing Materials, Milipitas, California. https://www.fireeye.com/content/dam/fireeye-www/global/en/products/pdfs/fireeye-hx-series.pdf.

Florida League of Cities. n.d. *About the League*. Accessed April 9, 2020. https://www.floridaleagueofcities.com/about-pages/about.

Florida State Legislature. 2022. "The 2023 Florida Statutes (including Special Session C)." *Online Sunshine*. July 1. Accessed January 12, 2024. http://www.leg.state.fl.us/statutes/index.cfm?App_mode=Display_Statute&Search_String=&URL=0200-0299/0282/Sections/0282.3186.html.

ForeScout. 2016. "Best Practices for CounterACT® Deployment: Wired Post-Connect." Deployment Guide, San Jose, California.

Freed, Benjamin. 2019. "One year after Atlanta's ransomware attack, the city says it's transforming its technology." *State Scoop*. March 22. Accessed December 28, 2020. https://statescoop.com/one-year-after-atlantas-ransomware-attack-the-city-says-its-transforming-its-technology/.

Freedman, Linn. 2020a. "Ransomware: To Pay or Not to Pay." *SHRM*. December 15. Accessed January 7, 2020. https://www.shrm.org/resourcesandtools/hr-topics/technology/pages/ransomware-to-pay-or-not-to-pay.aspx.

———. 2020b. "Security Economy Episode 3: Got a Privacy Plan? GDPR, CCPA, and the Rise of Ransomware." *Battleship Security*. July 2. Accessed December 23, 2020. https://battleshipsecurity.com/blog/security-economy-episode-3-got-a-privacy-plan-gdpr-ccpa-and-the-rise-of-ransomware-with-linn-freedman.

Gartner. n.d. "Small and Midsize Business (SMP)." *Gartner Glossary*. Accessed November 4, 2020. https://www.gartner.com/en/information-technology/glossary/smbs-small-and-midsize-businesses.

Gatlan, Sergiu. 2020a. "Blackbaud sued in 23 class action lawsuits after ransomware attack." *Bleeping Computer.* November 3. Accessed December 5, 2020. https://www.bleepingcomputer. com/news/security/blackbaud-sued-in-23-class-action-lawsuits-after-ransomware-attack/.

———. 2020b. "UHS restores hospital systems after Ryuk ransomware attack." *Bleeping Computer.* October 30. Accessed November 3, 2020. https://www.bleepingcomputer.com/ news/security/uhs-restores-hospital-systems-after-ryuk-ransomware-attack.

Ghafur, S., S. Kristensen, K. Honeyford, G. Martin, A. Darzi, and P. Aylin. 2019. *A retrospective impact analysis of the WannaCry cyberattack on the NHS.* Analysis, NPJ Digital Medicine. https://doi.org/10.1038/s41746-019-0161-6.

Gibson, Steve, and Leo Laporte. 2019. "Exposed Cloud Databases." *Security Now.* July 2. Accessed July 10, 2019. https://www.grc.com/sn/sn-721.htm.

Goodin, Dan. 2020. "A Patient Dies After a Ransomware Attack Hits a Hospital." *Wired.* September 9. Accessed September 9, 2020. https://www.wired.com/ story/a-patient-dies-after-a-ransomware-attack-hits-a-hospital/.

———. 2013. "You're infected—if you want to see your data again, pay us $300 in Bitcoins." Ars Technicha, October 17. https://arstechnica.com/information-technology/2013/10/ youre-infected-if-you-want-to-see-your-data-again-pay-us-300-in-bitcoins/.

Google. n.d. *BeyondCorp.* Accessed Febrary 8, 2020. https://cloud.google.com/beyondcorp.

Greenberg, Andy. 2020. "A Tesla Employee Thwarted an Alleged Ransomware Plot." *Wired.* August 27. Accessed August 27, 2020. https://www.wired.com/story/ tesla-ransomware-insider-hack-attempt/.

Gumbel, Andrew. 2011. "Kidnap makes an ass of Italy's ransom law." *Independent.* October 22. Accessed March 19, 2021. https://www.independent.co.uk/news/kidnap-makes-an-ass-of-italy-s-ransom-law-1144285.html.

Gupta, Aditya. 2017. "Understanding ShellExecute function and it's application to open a list of URLs present in a file using C++ code." *Geeks for Geeks.* July 7. Accessed July 26, 2020. https://www.geeksforgeeks.org/ understanding-shellexecute-function-application-open-list-urls-present-file-using-c-code/.

Hanslovan, Kyle. 2020. Huntress Labs, August 14.

Hanslovan, Kyle, and Chris Bisnett. 2020. "hack_it 2020." Online: Huntress Labs. https://cos. huntresslabs.com/hack_it_2020.

Hicock, Robyn. 2016. *Microsoft Password Guidance.* Microsoft. https://www.microsoft.com/ en-us/research/wp-content/uploads/2016/06/Microsoft_Password_Guidance-1.pdf.

Huffman, Bart W., Michael J. Lowell, Wendell J. Bartnick, and Julianne K. Nowicki. n.d. *Is Paying a Ransom to Stop a Ransomware Attack Illegal?* White paper, Reed Smith. https://www.sxsw. com/wp-content/uploads/2018/03/Legality-of-Paying-Ransom-FINAL-2018.1.19.pdf.

HYAS Intel Team. 2021. "Inside Ryuk Crime [Crypto] Ledger & Asian Crypto Traders." *HYAS.* January 7. Accessed January 20, January. https://www.hyas.com/blog/ inside-ryuk-crypto-ledger-asian-crypto-traders.

Ilascu, Ionut. 2021. "China's APT hackers move to ransomware attacks." *Bleeping Computer.* January 4. Accessed February 10, 2021. https://www.bleepingcomputer.com/news/security/ chinas-apt-hackers-move-to-ransomware-attacks/.

Intel 471. 2020. "Ransomware-as-a-service: The pandemic within a pandemic." *Intel 471.* November 16. Accessed November 24, 2020. https://intel471.com/blog/ ransomware-as-a-service-2020-ryuk-maze-revil-egregor-doppelpaymer/.

Jha, Somesh. 2021. "IntentOnBan, IndiaToGiveTransitionTimeToCryptoInvestors—BQExclusive." *Bloomberg Quint.* February 11. Accessed February 19, 2021. https://www.bloombergquint. com/business/intent-on-ban-india-to-give-transition-time-to-crypto-investors-bq-exclusive.

Johnson, O'Ryan. 2019. "'This Can't Be Happening': One MSP's Harrowing Ransomware Story." *CRN.* September 11. Accessed September 25, 2020. https://www.crn.com/news/ security/-this-can-t-be-happening-one-msp-s-harrowing-ransomware-story.

Kennelly, Jeremy, Kimberly Goody, and Joshua Shilko. 2020. "Navigating the MAZE: Tactics, Techniques and Procedures Associated With MAZE Ransomware Incidents." *FireEye Threat*

Research. May 7. Accessed September 20, 2020. https://www.fireeye.com/blog/threat-research/2020/05/tactics-techniques-procedures-associated-with-maze-ransomware-incidents.html.

KnowBe4. n.d. *AIDS Trojan or PC Cyborg Ransomware*. Accessed February 12, 2021. https://www.knowbe4.com/aids-trojan.

Krebs, Brian. 2019. "Ransomware Bites 400 Veterinary Hospitals." *Krebs on Security*. November 2019. Accessed November 8, 2020. https://krebsonsecurity.com/2019/11/ransomware-bites-400-veterinary-hospitals/.

———. 2020a. "Ransomware Group Turns to Facebook Ads." *Krebs on Security*. November 10. Accessed November 13, 2020. https://krebsonsecurity.com/2020/11/ransomware-group-turns-to-facebook-ads/.

———. 2020b. "The Hidden Cost of Ransomware: Wholesale Password Theft." *Krebs on Security*. January 6. Accessed February 29, 2020. https://krebsonsecurity.com/2020/01/the-hidden-cost-of-ransomware-wholesale-password-theft/.

———. 2020c. "Why Paying to Delete Stolen Data is Bonkers." *Krebs on Security*. November 4. Accessed January 10, 2021. https://krebsonsecurity.com/2020/11/why-paying-to-delete-stolen-data-is-bonkers/.

Kudale, Jack, interview by Douglas Brush. 2020. *#092—Jack Kudale: You Gotta Have More Cowbell* (June 22). https://cybersecurityinterviews.com/episodes/092-jack-kudale-you-gotta-have-more-cowbell/.

LaCroix, Kevin M. 2019. "Guest Post: Ransomeware's Dirty Little Secret: Most Corporate Victims Pay." *The D&O Diary*. January 31. Accessed March 12, 2020. https://www.dandodiary.com/2019/01/articles/cyber-liability/guest-post-ransomewares-dirty-little-secret-corporate-victims-pay/.

Langlois, Philippe. 2020. *2020 Data Breach Investigations Report*. Verizon. https://enterprise.verizon.com/resources/reports/dbir/.

Law Insider. n.d. "Sample Cyber Insurance Policy." *Law insider*. Accessed December 29, 2020. https://www.lawinsider.com/documents/2waqJ54HgoX.

Lemos, Robert. 2020. "'Act of War' Clause Could Nix Cyber Insurance Payouts." *Dark Reading*. October 29. Accessed December 31, 2020. https://www.darkreading.com/attacks-breaches/act-of-war-clause-could-nix-cyber-insurance-payouts/d/d-id/1339317.

Leyden, John. 2016. "When you've paid the ransom but you don't get your data back." *The Register*. September 7. Accessed January 12, 2021. https://www.theregister.com/2016/09/07/uk_ransomware_victim_survey.

Lighter Capital. n.d. *What Are Unit Economics and Why Are They Important in Early Stage Startups?* Accessed February 18, 2020. https://www.lightercapital.com/blog/what-are-unit-economics.

Lockheed Marten. n.d. "The Cyber Kill Chain." *lockheedmarten.com*. Accessed August 28, 2020. https://www.lockheedmartin.com/en-us/capabilities/cyber/cyber-kill-chain.html.

Lopatto, Elizabeth. 2019. "How Bitcoin Grew Up and Became Big Money." *The Verge*, January 13. https://www.theverge.com/2019/1/3/18166096/bitcoin-blockchain-code-currency-money-genesis-block-silk-road-mt-gox.

Lovejoy, Kris. 2020. "Ransomware: to pay or not to pay?" *EY*. January 10. Accessed January 10, 2021. https://www.ey.com/en_us/consulting/ransomware-to-pay-or-not-to-pay.

LR, Andrew, and Douglas AO. 2018. *Bitcoin Investigations: Evolving Methodologies and Case Studies*. Journal of Forensic Research. https://doi.org/10.4172/2157-7145.100042.

McAfee. 2020. *McAfee Network Security Platform 9.1.x Product Guide*. September 29. Accessed January 3, 2021. https://docs.mcafee.com/bundle/network-security-platform-9.1.x-product-guide/page/GUID-D8A9ECFD-CEEF-4157-8A0F-0DE3A5EF973A.html.

Meiklejohn, Sarah, Marjori Pomarole, Grant Jordan, Kirill Levchenko, Damon McCoy, Geoffrey M. Voelker, and Stefan Savage. 2013. *A Fistful of Bitcoins: Characterizing Payments Among Men with No Names*. Barcelona, Spain: ACM. https://doi.org/10.1145/2504730.2504747.

Melendez, Steven. 2016. "Ransomware Attacks Are Still On The Rise, Experts Warn." *Fast Company*. June 1. Accessed September 24, 2020. https://www.fastcompany.com/3060487/ransomware-attacks-are-still-on-the-rise-experts-warn.

Microsoft. n.d. *Ransomware detection and recovering your files.* Accessed February 17, 2021. https://support.microsoft.com/en-us/office/ransomware-detection-and-recovering-your-files-0d90ec50-6bfd-40f4-acc7-b8c12c73637f.

MITRE Corporation. n.d.-a "Xbash." *MITRE ATT&CK.* Accessed September 18, 2020. https://attack.mitre.org/software/S0341/.

———. n.d.-b "Disk Wipe: Disk Content Wipe." *MITRE ATT&CK.* Accessed September 3, 2020. https://attack.mitre.org/techniques/T1561/001/.

———. n.d.-c "Impact." *MITRE ATT&CK.* Accessed September 13, 2020. https://attack.mitre.org/tactics/TA0040/.

Morgan, Steve. 2020. "Cybercrime To Cost The World $10.5 Trillion Annually By 2025." Cybercrime Magazine, November 13. https://cybersecurityventures.com/hackerpocalypse-cybercrime-report-2016/.

Morning Consult + IBM Security. 2019. *Local Government Ransomware Study.* Survey, IBM. https://www.ibm.com/downloads/cas/MKPQVOL6.

Muncaster, Phil. 2020. "US Biz Closes Doors After Ransomware Attack." *Info Security.* January 7. Accessed December 30, 2020. https://www.infosecurity-magazine.com/news/us-biz-closes-doors-after.

Munshaw, Jon, and Joe Marshall. 2019. "What you — and your company — should know about cyber insurance ." *Cisco Talos.* August 20. Accessed May 22, 2020. https://blog.talosintelligence.com/2019/08/cyber-insurance-FAQs.html.

Munshaw, Jonathan. 2019. "Should governments pay extortion payments after a ransomware attack? ." *Cisco Talos.* July 11. Accessed January 6, 2021. https://blog.talosintelligence.com/2019/07/ransomware-extortion-roundtable-government-payments.html.

Nakamoto, Satoshi. 2008. "Bitcoin: A Peer-to-Peer Electronic Cash System." https://bitcoin.org/bitcoin.pdf.

National Conference of State Legislatures. 2020. "Cybersecurity Legislation 2020." *NCSL.* September 13. Accessed December 27, 2020. https://www.ncsl.org/research/telecommunications-and-information-technology/cybersecurity-legislation-2020.aspx.

Nelson, Tim, Andrew D. Ferguson, Michael J.G. Scheer, and Shriram Krishnamurthi. 2014. "Tierless Programming and Reasoning for Software-Defined Networks." *11th {USENIX} Symposium on Networked Systems Design and Implementation* 519–531. https://www.usenix.org/conference/nsdi14/technical-sessions/presentation/nelson.

NIST Computer Security Resource Center. 2020. "SP 800-207." *Zero Trust Architecture.* August. Accessed August 15, 2020. https://csrc.nist.gov/publications/detail/sp/800-207/final.

No More Ransom. n.d. "Ransomware Q&A." *No More Ransom.* Accessed January 11, 2021. https://www.nomoreransom.org/en/ransomware-qa.html.

North Carolina General Assembly. 2021. Accessed January 12, 2024. https://www.ncleg.gov/EnactedLegislation/Statutes/PDF/BySection/Chapter_143/GS_143-800.pdf.

OFAC. n.d. *Sanctions List Search.* Accessed December 11, 2020. https://sanctionssearch.ofac.treas.gov/.

OFAC_Feedback@treasury.gov, interview by Harry Halikias. 2020. (December 22).

Osborne, Charlie. 2020. "Logistics giant Toll Group hit by ransomware for the second time in three months." *ZDNet.* May 6. Accessed May 22, 2020. https://www.zdnet.com/article/transport-logistics-firm-toll-group-hit-by-ransomware-for-the-second-time-in-three-months/.

Palmer, Danny. 2019a. "Ransomware: Big paydays and little chance of getting caught means boom time for crooks." *ZDNet.* November 29. Accessed December 8, 2019. https://www.zdnet.com/article/ransomware-big-paydays-and-little-chance-of-getting-caught-means-boom-time-for-crooks/.

———. 2019b. "Ransomware: Cyber-insurance payouts are adding to the problem, warn security experts." *ZDNet.* September 17. Accessed December 30, 2020. https://www.zdnet.com/article/ransomware-cyber-insurance-payouts-are-adding-to-the-problem-warn-security-experts/.

Pamela. 2020. "Tracing Ransomware: CipherTrace Helps McAfee Follow NetWalker Funds." *CipherTrace.* August 6. Accessed December 27, 2020. https://ciphertrace.com/tracing-ransomware-ciphertrace-helps-mcafee-follow-netwalker-funds/.

Plesco, Ron, and Stacy Shelhorse. 2020. "When a threat actor strikes: Legal considerations and challenges in a ransomware attack." *DLA Piper.* December 21. Accessed January 12, 2021. https://www.dlapiper.com/en/us/insights/publications/2020/12/understanding-ransomware-stratagems.

Puodzius, Cassius. 2016. "How encryption molded crypto-ransomware." *We Live Security by ESET.* September 13. Accessed September 24, 2020. https://www.welivesecurity.com/2016/09/13/how-encryption-molded-crypto-ransomware/.

PYMNTS. 2018. "Forget Bitcoin — Privacy Tokens A Favorite Of The Bad Guys." *pymnts.com.* January 3. Accessed January 21, 2020. https://www.pymnts.com/news/blockchain-distributed-ledger/2018/bitcoin-privacy-coins-tokens-zcash-monero-ethereum/.

Ramachandran, Seetha, Nolan Goldberg, and Hena Vora. 2020. "Regulatory Crackdown on Ransomware." *The National Law Review.* December 15. Accessed December 23, 2020. https://www.natlawreview.com/article/regulatory-crackdown-ransomware.

Raver, Carrie Marie. 2019. "A Ransomware Attack Could Devastate Your Company. Will Your Insurance Cover It?" *The National Law Review.* November 5. Accessed December 31, 2020. https://www.natlawreview.com/article/ransomware-attack-could-devastate-your-company-will-your-insurance-cover-it.

Rivero Lopez, Marc. 2020. "Tales From the Trenches; a Lockbit Ransomware Story." *McAfee.* April 30. Accessed July 21, 2020. https://www.mcafee.com/blogs/other-blogs/mcafee-labs/tales-from-the-trenches-a-lockbit-ransomware-story.

Sarah. 2017. "Spotlight on ransomware: Ransomware encryption methods." *Emsisoft Blog.* June 21. Accessed September 2020, 23. https://blog.emsisoft.com/en/27649/ransomware-encryption-methods/.

Satariano, Adam, and Nicole Perloth. 2019. "Big Companies Thought Insurance Covered a Cyberattack. They May Be Wrong." *The New York Times.* April 15. Accessed January 2, 2020. https://www.nytimes.com/2019/04/15/technology/cyberinsurance-notpetya-attack.html.

Sayce, Scott. 2024. "3 trends set to drive cyberattacks and ransomware in 2024." *World Economic Forum.* February 22. Accessed March 24, 2024. https://www.weforum.org/agenda/2024/02/3-trends-ransomware-2024.

Schwartz, Mathew. 2020. "Russia's Cybercrime Rule Reminder: Never Hack Russians." *Bank Info Security.* March 27. Accessed April 24, 2020. https://www.bankinfosecurity.com/blogs/russias-cybercrime-rule-reminder-never-hack-russians-p-2888.

Seals, Tara. 2018. "One Year After WannaCry: A Fundamentally Changed Threat Landscape." *Threat Post.* May 17. Accessed November 8, 2020. https://threatpost.com/one-year-after-wannacry-a-fundamentally-changed-threat-landscape/132047/.

Security Magazine. 2020. "Fairfax County Public Schools hit by Maze ransomware." *Seurity Magazine.* September 15. Accessed September 23, 2020. https://www.securitymagazine.com/articles/93354-fairfax-county-public-schools-hit-by-maze-ransomware.

SentinelOne. 2020. ""EvilQuest" Rolls Ransomware, Spyware & Data Theft Into One." *SentinelOne.* July 8. Accessed September 22, 2020. https://www.sentinelone.com/blog/evilquest-a-new-macos-malware-rolls-ransomware-spyware-and-data-theft-into-one/.

———. 2018. "Meterpreter: The Advanced and Powerful Metasploit Payload." *SentinelOne.* September 6. Accessed August 7, 2020. https://www.sentinelone.com/blog/meterpreter-advanced-powerful-metasploit-payload/.

———. 2019. "Ransomware as a Service | What are Cryptonite, Recoil and Ghostly Locker?" *SentinelOne.* December 16. Accessed June 28, 2020. https://www.sentinelone.com/blog/ransomware-as-a-service-what-are-cryptonite-recoil-and-ghostly-locker.

Simmons + Simmons. 2018. *The legality of cyber extortion payments.* December 14. Accessed December 21, 2020. https://www.simmons-simmons.com/en/publications/ck0ahwpb0ncm30b369kyk8e7o/131218-the-legality-of-cyber-extortion-payments.

Spadafora, Anthony. 2020. "FBI: Over $140 million handed over to ransomware attackers." *Tech Radar Pro.* February 28. Accessed September 22, 2020. https://www.techradar.com/news/fbi-over-dollar140-million-handed-over-to-ransomware-attackers.

Stender, Walter W., and Evans Walker. 1974a. "The National Personnel Records Center Fire: A Study in Disaster." *The American Archivist* 521–549. https://meridian.allenpress.com/american-archivist/article/37/4/521/22794/The-National-Personnel-Records-Center-Fire-A-Study.

———. 1974b. "The National Personnel Records Center Fire: A Study in Disaster." *The American Archivist* 521–549. https://www.archives.gov/files/st-louis/military-personnel/nprc-fire.pdf.

Taylor, Mike, Mitch Thornton, and Kaitlin Smith. 2020. "Software developed by SMU stops ransomware attacks." *SMU Research News.* May 13. Accessed November 14, 2020. https://blog.smu.edu/research/2020/05/13/software-developed-by-smu-stops-ransomware-attacks/.

Texas Department of Information Resources. 2019. "Update on Texas Local Government Ransomware Attack ." *Texas Department of Information Resources.* September 19. Accessed November 6, 2020. https://dir.texas.gov/View-About-DIR/Article-Detail.aspx?id=213.

The New York State Senate. n.d. "New York State Senator Phil Boylse." Accessed February 13, 2020. https://www.nysenate.gov/senators/phil-boyle/about.

Thompson, Garrett. 2020. "REvil RaaS Means Business." September 28. Accessed October 3, 2020. https://www.binarydefense.com/threat_watch/revil-raas-means-business/.

Thornton, Mitchell, Michael Taylor, and Kaitlin Smith. 2020. Detecting Malicious Software Using Sensors. United States Patent Application number: 15/812663. September 3.

Tiwari, Ravikant. 2020. "Taking Deep Dive into Sodinokibi Ransomware." *Acronis Case.* Accessed September 2020, 2020. https://www.acronis.com/en-us/articles/sodinokibi-ransomware/.

UK Natoinal Audit Office Comptroller and Auditor General. 2018. "Investigation: WannaCry cyber attack and the NHS ." *National Audit Office.* April 25. Accessed September 20, 2020. https://www.nao.org.uk/wp-content/uploads/2017/10/Investigation-WannaCry-cyber-attack-and-the-NHS.pdf.

Ulrich, Johannes. 2019. "Another Word Maldoc; Snatch Ransomware; Ryuk Decryptor Fail; Sysmon DNS Rules." *SANS Internet Storm Center.* December 10. Accessed December 10, 2019. https://isc.sans.edu/podcastdetail.html?id=6782.

US CISA. 2020. "Advanced Persistent Threat Activity Exploiting Managed Service Providers." *Alert (TA18-276B).* October 18. Accessed November 5, 2020. https://us-cert.cisa.gov/ncas/alerts/TA18-276B.

US CISA, FBI, and DHS. 2020. "Ransomware Activity Targeting the Healthcare and Public Health Sector." Advisory, Washington DC. https://us-cert.cisa.gov/sites/default/files/publications/AA20-302A_Ransomware%20_Activity_Targeting_the_Healthcare_and_Public_Health_Sector.pdf.

US Department of Health and Human Services. n.d. *Fact Sheet: Ransomware and HIPAA.* Advisory, Washington DC: HHS.gov. https://www.hhs.gov/sites/default/files/RansomwareFactSheet.pdf.

———. 2020a. "Maze Ransomware." *HHS Cybersecurity Program.* June 4. Accessed August 7, 2020. https://www.hhs.gov/sites/default/files/maze-ransomware.pdf.

———. 2020b. "TrickBot, Ryuk, and the HPH Sector." *HHS Cybersecurity Program.* November 12. Accessed November 14, 2020. https://www.hhs.gov/sites/default/files/trickbot-ryuk-and-the-hph-sector.pdf.

US Department of Justice. 2020a. *Cryptocurrency Enforcement Framework.* Washington D.C.: US Department of Justice. https://www.justice.gov/cryptoreport.

———. 2018. "North Korean Regime-Backed Programmer Charged With Conspiracy to Conduct Multiple Cyber Attacks and Intrusions." *Office of Public Affairs.* September 8. Accessed November 21, 2020. https://www.justice.gov/opa/pr/north-korean-regime-backed-programmer-charged-conspiracy-conduct-multiple-cyber-attacks-and.

———. 2020b. "Six Russian GRU Officers Charged in Connection with Worldwide Deployment of Destructive Malware and Other Disruptive Actions in Cyberspace." *Office of Public Affairs.* October 19. Accessed October 20, 2020. https://www.justice.gov/opa/pr/six-russian-gru-officers-charged-connection-worldwide-deployment-destructive-malware-and.

———. 2023. "U.S. Department of Justice Disrupts Hive Ransomware Variant." *Office of Public Affairs.* January 26. Accessed January 29, 2023. https://www.justice.gov/opa/pr/us-department-justice-disrupts-hive-ransomware-variant.

———. 2020c. "UK National Sentenced to Prison for Role in "The Dark Overlord" Hacking Group." *Office of Public Affairs.* September 21. Accessed December 8, 2020. https://www.justice.gov/opa/pr/uk-national-sentenced-prison-role-dark-overlord-hacking-group.

US Department of The Treasury. 2002. *Frequently Asked Questions.* September 10. Accessed December 10, 2020. https://home.treasury.gov/policy-issues/financial-sanctions/faqs/topic/1501.

———. n.d.-a *OFAC Frequently Requested.* Accessed December 20, 2020. https://home.treasury.gov/footer/freedom-of-information-act/electronic-read-room/ofac-frequently-requested.

———. n.d.-b *Program Tag Definitions for OFAC Sanctions Lists .* Accessed December 10, 2020. https://home.treasury.gov/policy-issues/financial-sanctions/specially-designated-nationals-list-sdn-list/program-tag-definitions-for-ofac-sanctions-lists.

———. n.d.-c *Sanctions List Updates.* Accessed December 11, 2020. https://home.treasury.gov/policy-issues/financial-sanctions/recent-actions/1316.

US Department of The Treasury, Office of Foreign Assets Control (OFAC). 2020. *Advisory on Potential Sanctions Risks for Facilitating Ransomware Payments.* Advisory, Washington DC: Department of The Treasury. https://home.treasury.gov/system/files/126/ofac_ransomware_advisory_10012020_1.pdf.

US Department of Veteran Affairs. n.d. "Reconstruct military records destroyed in NPRC fire." *VA. gov.* Accessed November 1, 2020. https://www.va.gov/records/get-military-service-records/reconstruct-records/.

US District Court, Eastern District of MO St. Louis. 2017. "United States of America v. Nathan Wyatt." Court document, St. Louis, MO. https://www.justice.gov/opa/press-release/file/1227441/download.

US Securities and Exchange Commission. 2018. "Commission Statement and Guidance on Public Company Cybersecurity Disclosures." Advisory, Washington D.C. https://www.sec.gov/rules/interp/2018/33-10459.pdf.

Web Titan. 2019. "How Much Money Did WannaCry Make?" *Web Titan.* December 2. Accessed September 9, 2020. https://www.webtitan.com/blog/how-much-money-did-wannacry-make.

Wellerman, Zak, and Tyler Morning. 2020. "Texas School District Forks Over $50K in Ransomware Attack." *Government Technology.* July 31. Accessed November 14, 2020. https://www.gov-tech.com/security/Texas-School-District-to-Fork-Over-50K-in-Ransomware-Attack.html.

Widup, Suzanne. 2019. "2019 Verizon Data Breach Investigations Report ." *Verizon.* October. Accessed November 9, 2020. https://www.nist.gov/system/files/documents/2019/10/16/1-2-dbir-widup.pdf.

Wilkie Compliance. n.d. *To Whom Do Sanctions Apply.* Accessed December 21, 2020. https://complianceconcourse.willkie.com/resources/sanctions-us-who-must-comply-with-sanctions.

Worlock, Charlotte. 2020. "Comparing US and EU Approaches to Paying Cyber Ransoms." *Atheria Law.* January 23. Accessed December 9, 2020. https://atherialaw.com/what_happening/comparing-us-and-eu-approaches-to-paying-cyber-ransoms/.

Xiao, Claud, Cong Zheng, and Xingyu Jin. 2017. "Xbash Combines Botnet, Ransomware, Coinmining in Worm that Targets Linux and Windows." *Palo Alto Unit 42.* September 17. Accessed September 20, 2020. https://unit42.paloaltonetworks.com/unit42-xbash-combines-botnet-ransomware-coinmining-worm-targets-linux-windows/.

Young, Adam, and Moti Yung. 1996. "Cryptovirology: extortion-based security threats and coun-termeasures." *IEEE Symposium on Security & Privacy* 129–141. https://www.ieee-security.org/TC/SP2020/tot-papers/young-1996.pdf.

Zelonis, Josh. 2019. "https://go.forrester.com/blogs/unconventional-wis-dom-explore-paying-the-ransom-in-parallel-with-other-recovery-options/." *Forrester.* June 4. Accessed December 27, 2020. https://go.forrester.com/blogs/unconventional-wisdom-explore-paying-the-ransom-in-parallel-with-other-recovery-options/.